METAPHYSICS AND THE PHILOSOPHY OF SCIENCE

Metaphysics and the Philosophy of Science

NEW ESSAYS

Edited by Matthew H. Slater

and Zanja Yudell

OXFORD
UNIVERSITY PRESS

OXFORD
UNIVERSITY PRESS

Oxford University Press is a department of the University of Oxford. It furthers the University's objective of excellence in research, scholarship, and education by publishing worldwide. Oxford is a registered trade mark of Oxford University Press in the UK and certain other countries.

Published in the United States of America by Oxford University Press
198 Madison Avenue, New York, NY 10016, United States of America.

Library of Congress Cataloging-in-Publication Data
Names: Slater, Matthew H., 1977– editor. | Yudell, Zanja, 1975– editor.
Title: Metaphysics and the philosophy of science : new essays /
edited by Matthew Slater and Zanja Yudell.
Description: New York, NY : Oxford University Press, [2017] |
Includes bibliographical references and index.
Identifiers: LCCN 2016029883 | ISBN 9780199363209 (cloth : alk. paper)
Subjects: LCSH: Science—Philosophy. | Metaphysics. | Philosophy of nature.
Classification: LCC Q175 .M541655 2017 | DDC 501—dc23
LC record available at https://lccn.loc.gov/2016029883

9 8 7 6 5 4 3 2 1

Printed by Sheridan Books, Inc., United States of America

Contents

Author Bios

Katherine Brading is Professor of Philosophy at the University of Notre Dame. She works on topics in history and philosophy of physics from the seventeenth century to the present, including spacetime theory, symmetries and conservation laws, structuralist approaches to physics, and matter theory. Her current project concerns eighteenth century natural philosophy in the wake of Newton's *Principia*, with a particular focus on Émilie Du Châtelet.

Jenann Ismael is a philosopher of physics. She has written on probability, time, causation, modality, and quantum ontology. She has also engaged with issues in the philosophy of mind, applying the formal apparatus associated with symmetry to understanding interactions among perspectives, self-location, and consciousness. Recently she has been thinking about the relationship between physics and free will. She has been a professor at the University of Arizona since 1996 and held fellowships at Stanford, the National Humanities Center, and a QEII fellowship from the Australian Research Council.

James Ladyman studied pure mathematics and philosophy at the University of York, and then took a masters in history and philosophy of science and mathematics at King's College London. He completed his PhD, on the semantic approach to scientific theories and structural realism, under the supervision of Steven French at the University of Leeds in 1997. He has been assistant, deputy and co-editor of the British Journal for the Philosophy of Science and honorary secretary of the

British Society for the Philosophy of Science. He is professor of philosophy at the University of Bristol.

Juha Saatsi is an Associate Professor of Philosophy at University of Leeds. He works on a variety of topics in the philosophy of science, including scientific realism and theories of explanation.

Matthew H. Slater is an Associate Professor of Philosophy at Bucknell University, having received his PhD from Columbia University. He is the author of *Are Species Real?* (Palgrave, 2013) and has co-edited such volumes as *Carving Nature at its Joints, The Environment*, and *Reference and Referring* (MIT Press) and writes on issues in the philosophy of science and metaphysics.

P. Kyle Stanford is professor and chair of the Department of Logic and Philosophy of Science at the University of California, Irvine. He is the author of *Exceeding Our Grasp: Science, History, and the Problem of Unconceived Alternatives* and many further articles concerned with scientific realism and instrumentalism, as well as the philosophy of biology, the history of modern philosophy, and the philosophy of language. He is currently working on evolutionary explanations of human moral psychology and of cognition more generally.

Michael Strevens is a professor of Philosophy at New York University. He was born and raised in New Zealand. He moved to the United States in 1991 to undertake a PhD at Rutgers University; currently, he teaches philosophy of science at New York University. His academic work covers topics such as understanding, complexity, causation, and the social structure of science, as well as the philosophical applications of cognitive science.

Martin Thomson-Jones is a professor of Philosophy at Oberlin College. Before Oberlin, he taught at Princeton and then at the University of California, Berkeley. His current research focuses on a cluster of questions about representation in the sciences, including questions about the nature of models and modeling, and about the connections between scientific representation and "ordinary" fiction. He has also worked in the philosophy of physics and in related areas of metaphysics.

C. Kenneth Waters has held the Canada Research Chair in Logic and Philosophy of Science at the University of Calgary since 2014. Previously, he served as Director for the Minnesota Center for Philosophy of Science. His research examines the nature of knowledge in the context of scientific practices and the conditions that make this knowledge possible. Much of his work examines biological sciences and the way biologists seek to investigate, manipulate, and understand life.

James Woodward is distinguished professor in the Department of History and Philosophy of Science at the University of Pittsburgh. Prior to 2010, he was the J.O. and Juliette Koepfli Professor at the California Institute of Technology. Much of his work, including his 2003 book, *Making Things Happen: A Theory of Causal Explanation*, explores issues having to do with causation and explanation, but he also maintains an interest in scientific methodology more generally.

Zanja Yudell is an associate professor of Philosophy at California State University at Chico, having received his PhD from Columbia University. He works on topics in the philosophy of science, especially laws of nature and explanation.

METAPHYSICS AND THE PHILOSOPHY OF SCIENCE

Introduction

Zanja Yudell

HAVE WE PHILOSOPHERS of science come to bury metaphysics or to praise it? More than one observer has noted that the history of philosophy has followed a cycle in which metaphysics repeatedly falls out of favor, only to return again, characterizing this cycle using metaphorical language of death, burial, and resurrection (e.g., Callendar 2011, 35; Lowe 2011, 102–5; Price 2009, 322). Within the Anglo-American tradition, there is a fairly well-known story about a recent iteration of the cycle in the fortunes of metaphysics. The story begins in the early twentieth century with the empiricist rejection of metaphysics. As described by Ayer (1936), the logical positivists' and empiricists' rejection of metaphysics ultimately derived from the verifiability theory of meaning: strictly speaking, metaphysical claims are not verifiable and hence meaningless. By contrast, scientific claims offer themselves as clear paradigms of verifiability and hence meaningfulness. They aimed to make philosophy rise to the standard of science, and an essential component of that project was to expunge metaphysics from philosophy (Creath 2011). Thus, a line in the sand had been drawn: to claim the mantle of science (and scientific philosophy) was to deny that metaphysics could be profitably pursued.

As the story continues, Quine restores metaphysics in the middle of the century by denying the analytic/synthetic distinction, thus erasing the strict boundary between science and philosophy (whether such a boundary was policed by the logical empiricists or not) and eliminating verifiability as a standard by which the meaninglessness of metaphysical claims could be established. Metaphysics could now be respectably pursued again, but not as the free-floating, a priori exploration of the days of yore.

Quinean metaphysics is *naturalized*, which means in part that it must be responsive to the natural sciences. However, the continuity between philosophy and science that allows metaphysics back into polite company requires that metaphysics must submit its claims to the tribunal of science, if not experience.

Many philosophers have acquiesced to this requirement, and the decades since Quine have seen a flowering of metaphysical work in the Anglo-American tradition that is ostensibly naturalistic. Quine's naturalism favored a rather austere vision of metaphysics (Quine 1948, 23), but others claiming the mantle of naturalism have endorsed somewhat more extravagant landscapes (e.g., Dennett 1991). So-called analytic metaphysics can be yet more extravagant, placing center stage the modality (Lowe 2011) that Quine rejected (e.g., Quine 1960), and employing concepts such as atomless gunk that are not applicable to the *actual* world, according to our best science. Nonetheless, although oriented toward the a priori, many such metaphysicians seem to have accepted Quine's bargain in some measure, endorsing a continuity between metaphysics and science, if not in subject matter, then in methodology (Paul 2012; Sider 2008, 6). It is not uncommon to see metaphysicians endorse the standard virtues of scientific theories (simplicity, explanatory power, consistency, fruitfulness, and even empirical adequacy; see, e.g., Paul 2012) as applicable tests for metaphysical views as well.

The pendulum may now be swinging back away from metaphysics: Callendar (2011, 34–35) remarks that the type of self-examination that has recently sprung up in metaphysics and is known as metametaphysics (see Chalmers, Manley, and Wasserman 2009) is "[never] a good sign for a field." Kuhn observed that one of the signs of a crisis in science is when practitioners begin asking methodological questions (Kuhn [1962] 1996, 88). We can all agree that analytic metaphysics is not a science, yet it certainly has many of the institutional characteristics of Kuhnian science. Moreover, prominent philosophers of science have recently attacked metaphysics (Ladyman et al. 2007, chap. 1) and stirred up some dust (Dorr 2010). Nonetheless, it is a bit too early to ring the death knell: after all, a Kuhnian crisis can resolve in a number of ways that result in the continuation of "normal science."

Whatever its errors or oversimplifications, this story is useful for highlighting how much weight many give to science in our current discussions about metaphysics. Those who reject metaphysics outright do so in the name of science; others grudgingly give it room at the table as long as it respects the authority of science; but even many of those who assert the autonomy or primacy of metaphysics hitch their defense of its legitimacy to its methodological continuity or similarity with science.

If a proper understanding of the status of metaphysics involves taking a position on its relation to science, such an understanding requires a conception of science. Presumably, the more sophisticated the conception of science, the better the understanding. Some have accused contemporary analytic metaphysicians of having an outmoded picture of science that contributes to misguided and empty metaphysical theorizing (Ladyman et al. 2007, chap. 1; Ladyman, this volume, chap. 7). To whatever extent that accusation is true, it should be uncontroversial that contemporary philosophers of science have views of science that are more deeply connected to the practices of science. Philosophers of science, then, should have much to contribute to the project of making sense of metaphysics. In this volume, we have assembled essays by philosophers of science who have reflected on the relation between metaphysics and science by coming from the "science end," so to speak. A guiding idea behind this book has been that the perspective of philosophers of science will add a useful contribution to the current discussions about the status of metaphysics and allow readers to enter the debates from the variety of perspectives found within this community of thinkers.

Consider how Post-Quinean metaphysics has been undertaken by analytic metaphysicians as opposed to philosophers of science. If you look through the tables of contents of recent texts in metaphysics or collections of essays in metaphysics, you will see, for example, such topics as personal identity, free will, space and time, the nature of abstract entities, causation, modality, mereology, and ontological grounding (Sider, Hawthorne, and Zimmerman 2008; van Inwagen 2007). Many of these topics are, at least homonymically, traditional topics in metaphysics, although many of them are certainly now addressed by analytic metaphysicians in a more or less "Quinean" way, with explicit connections to contemporary science. And of course, some analytic metaphysicians explore these issues in ways that are arguably a bit less continuous with science (e.g., the aforementioned gunk).

If you look at philosophy of science texts and collections that purport to be about *general* philosophy of science, you will typically find a few topics that bear close relations to metaphysics: laws of nature, natural kinds, causation, and scientific realism (Balashov and Rosenberg 2002; Curd, Cover, and Pincock 2012). Laws of nature, natural kinds, and causation, although they may have a distinctive cast when discussed by philosophers of science, nonetheless have a fair amount of "crossover" with the discussions of such issues in metaphysics proper: metaphysics texts also give significant attention to these topics.

The truly *distinctive* metaphysical question within the philosophy of science for the past few decades has been scientific realism. It is common to distinguish the epistemic commitment of the realist from the metaphysical one (e.g., Psillos 1999, xvii), and sometimes from a semantic commitment as well. The metaphysical position is often expressed as the claim that the world investigated by the sciences exists independently of minds (Chakravartty 2011). Even van Fraassen, both an anti-realist and an opponent of (pre-Kantian) metaphysics (van Fraassen 1989, viii), accepts this claim (van Fraassen 1980, 38). But the terms assigned to this distinction are misleading: although van Fraassen may share the "metaphysical" commitment with the realist, he is not wrong to accuse realism of "inflationary metaphysics" (van Fraassen 1980, 73) resulting from its *epistemic* commitment. This second plank of scientific realism holds that science at least aims to deliver truths about the unobservable realm, if not that it actually delivers some such truths. To take seriously scientific statements as candidates for truth means to take seriously not just unobservable entities like quarks and genes, but also laws of nature, natural kinds, causes, and other furniture of the metaphysician's universe.

As scientific realism has evolved, some realists have sought a position with more modest commitments: entity realism (Hacking 1983), structural realism (Worrall 1989), and semirealism (Chakravartty 1998) are all attempts to selectively withdraw from endorsing everything imagined in our most successful scientific theories. But they all go beyond the world of experience, and in that sense go down the metaphysical road. Philosophy of science, in grappling with realism, thus apparently has no choice but to confront a metaphysical question. We might imagine that the metaphysics of scientific realism is the "gateway drug" that opens up young philosophers of science to other, more dangerous metaphysical questions.

Scientific realism is a thread that winds its way through at least half the papers in this volume explicitly. For example, in their own ways, Jenann Ismael (chap. 5) and Juha Saatsi (chap. 8) try to make sense of what kind of realism about science may be endorsed without opening the door too widely for more speculative metaphysics to follow in afterward. Ismael aims to provide a respectable empiricist understanding of modality and concludes that neither a pure instrumentalism nor realism can do the job. Saatsi focuses on philosophical methodology, arguing that the explanationist strategy for defending scientific realism does not carry over to the metaphysical realm. James Ladyman's contribution reaffirms the structural and rainforest realisms of Ladyman et al. (2007), while working out the consequences of these (and other) views for the metaphysics of composition. Ladyman takes those realisms to imply that there is no general answer to the question—what science reveals is different criteria in different domains.

Realism may be the most prominent metaphysical issue in general philosophy of science, but philosophy of science is not just general philosophy of science; it is also philosophy of the specific sciences. Philosophers of physics have engaged with many metaphysical questions, both traditional and novel, in the general project of making sense of physical theories (and experiments). Space and time are perennial subjects of metaphysics. Metaphysicians have not only used physics to answer the traditional questions, but those questions have been transformed by engagement with physics. For example, in chapter 1, Katherine Brading argues that metaphysical questions about the nature of time were able to be addressed empirically after Newton's *Principia*, although they did not suddenly cease to be metaphysical questions. Similarly, Ladyman, in chapter 7, discusses how the metaphysical question of special composition should be reconceived given contemporary physics. Kyle Stanford (chap. 6) calls such investigations of traditional metaphysical questions in light of scientific developments "scientistic metaphysics." But philosophers of physics have also been engaged in what he calls "metaphysics of science," which involves understanding the metaphysics inherent in scientific theories without seeking to determine whether the answers are correct descriptions of reality. Attempts to interpret quantum mechanics, give an ontology for field theories, or resolve the problem of the arrow of time might all be seen as metaphysics of science in Stanford's sense.

If metaphysics is more strongly associated with philosophy of physics than philosophy of biology or philosophy of the social sciences, it may be because one thinks of the relationship between metaphysics and science as primarily mediated by scientistic metaphysics. But one can undertake metaphysics of science with respect to any science. In this volume we have attempted to exhibit more of the relation between metaphysics and the special sciences than is commonly perceived. Matthew Slater, Kenneth Waters, and Michael Strevens all use examples from the life sciences to tease out novel insights about metaphysics and its relation to science. Slater examines three cases in which biology seems to dictate to metaphysics. In one of those cases, species essentialism, he agrees that the results of empirical work determine metaphysical claims to be false. But in two other cases, the thesis that species are individuals and Boyd's homeostatic property cluster proposal for analyzing natural kinds, he concludes that the metaphysical implications are not so easily drawn. Strevens argues that the success of science using "distributed ontologies" to explain complex systems implies a kind of ontological pluralism. Waters's chapter focuses on the nature of the gene and what contemporary genetics shows us about metaphysics. He concludes that a traditional metaphysics that pursues "the structure" of the world must be abandoned in favor of a new metaphysics that recognizes the inherent "messiness" of the world.

A common background as philosophers of science is perhaps the most significant unity to be found among the authors in this volume. This book is not a collection of essays by like-minded philosophers sharing a manifesto and a common vision of the proper pursuit of metaphysics. Nor is this volume, like *Scientific Metaphysics* (Ross, Ladyman, and Kincaid 2013), focused exclusively on the specific question of naturalizing metaphysics. Indeed, many of the authors here would call themselves naturalists, as would many philosophers of science today. But the questions they address are diverse, and collectively they give us a more general picture of the relationship between science and metaphysics.

Given the story I told earlier, one might naturally think that the collection of possible attitudes falls on a kind of spectrum ranging from the strict anti-metaphysics associated with the logical positivists to the extreme metaphysical imperialist who sees science flowing from the autonomous and fundamental activity of metaphysics, with the middle of the this spectrum occupied by attitudes according more or less priority to science or metaphysics. Although this spectrum is not a wholly useless way to think about things, a slightly more complex picture can better capture the available diversity of views concerning science and metaphysics, and help characterize something of the distinctive position held by philosophers of science, at least in this volume. Imagine a space with two axes. The horizontal axis represents the extent to which metaphysics depends on science, and the vertical axis represents the extent to which science depends on metaphysics. Thus, in the upper right are views on which the two are mutually interdependent; the lower left contains views on which they are largely independent of each other; on the lower right are the views on which metaphysics asymmetrically depends on science; and on the upper left we find the reverse. It would be taking this picture a bit too seriously to try to be precise about what "dependence" means here, but I think it is fair to say that in the relevant sense, a practice or activity *A* depends on *B* if the legitimacy or justification of *A* requires some degree of success in *B*. For example, in this sense of "dependence," applying a certain statistical test depends on choosing a random sample, and prescribing medication depends on having the proper license. So metaphysics depends on science to the extent that (some) metaphysical activity cannot be justified in the absence of specific scientific achievements. If one thinks, for example, that metaphysics is never justified, then we might class that view as a trivial case of metaphysics depending on science.

Perhaps unsurprisingly, the philosophers of science contributing to this volume take a range of opinions on the extent to which science depends on metaphysics and vice versa, but there are none who think that science largely depends on metaphysics but metaphysics is independent of science. In other words, none occupy the upper left quadrant—none of our authors are Cartesian rationalists. I think Brading is the author most clearly located in the upper right—she explicitly describes the interdependence of metaphysics and physics in her piece. I am tempted to place Martin Thomson-Jones's chapter in that corner as well, although it concerns the interdependence of metaphysics and the *philosophy* of science more than science, as discussed below. Indeed, nearly all of the essays in this volume belong on the right side of the space, as they argue in one way or another for the dependence of metaphysics on science. Fewer discuss precisely to what extent science depends on metaphysics, so one cannot always clearly locate them along the vertical axis. One clear exception is the dialogue by Woodward, in which the main character, Jim Woodward, argues for the independence of philosophical questions about the practice of science from metaphysics.

It is worth adding a few epicycles and equants to this picture. One is the well-known ambiguity in the word "science," which can refer to both an activity and the product of that activity. A similar ambiguity applies to "metaphysics." So, in charting the interdependencies of science and metaphysics, one might take care to distinguish whether the dependency flows from, say, the methodology of science to that of metaphysics, or from the conclusions of metaphysics to the practice of physics, and so on. Within this volume, authors explore a variety of different relations between these two aspects. For example, Strevens's chapter effectively argues that a certain scientific strategy for addressing complex systems, the construction of compositional models, is largely independent of any underlying fundamental ontology. Waters's chapter also focuses on the practice of science, but argues that substantial conclusions about metaphysical issues can be drawn from this practice. Ladyman similarly argues for a connection between science and metaphysics, although he argues that it is the methodology of metaphysics that must be constrained by the conclusions of physics. Finally, Saatsi's contribution focuses on the relationship between the practice of metaphysics and the practice of science, whether metaphysics can import the methods of science. His target is explanationism in metaphysics, which, like explanationism in science, seeks to justify inferences on the basis of their ability to explain the "phenomena."

A second issue that complicates our framework for categorizing views is the relationship between science and *philosophy* of science. As the title of this volume indicates, these essays do not simply address the relationship between metaphysics and science, but how metaphysics relates to the philosophy of science as well. The aforementioned ambiguity between practice and product applies here as well, and so in exploring the role of philosophy of science here, we may also conceive of it as an activity or the fruit of that activity. As a product, philosophy of science gives us images of science (both as product and activity) that we use in understanding how science relates to metaphysics. The theoretical virtues of science that metaphysicians adopt (or not) for their metaphysical theories are the fruit not of science but of philosophy of science. The interpretation of quantum mechanics that grounds (or not) a metaphysical rejection of the principle of the identity of indiscernibles is not simply a deliverance of physics, but a result of philosophical engagement with physics (although I grant that here the boundary between physics and philosophy is not always clear). It is not implausible to say that the work of bringing together metaphysics and science is always underpinned by a philosophy of science, with philosophy of science conceived of as product.

But our dichotomy also allows us to conceive of philosophy of science as an activity, and so we can ask how the activity of philosophy of science relates to metaphysics, which is a question that is less often considered. In this volume, Woodward and Thomson-Jones both ask whether *doing* philosophy of science depends on metaphysics, and they give different answers, albeit in different domains. Woodward presents his concerns in the form of a dialogue in which the character Jim Woodward expresses his frustration with a metaphysician who insists that he cannot understand causation, explanation, and related ideas without a notion of metaphysical grounding, arguing that we need do no more than understand how scientists discover causes and construct explanations. Thomson-Jones, by contrast, argues that there is at least one kind of question of interest to philosophers of science that cannot be properly answered without engaging with metaphysics: describing how models relate to the world.

One way in which the above framework oversimplifies things is in presenting all questions about the relationship between metaphysics and science as concerned with the dependence of each on the other. I think this simplification is useful and can have a broad enough scope if dependence is understood loosely enough, but there are certainly other important questions about this relationship that cannot be fit into this schema. One important question that has arisen many times over the years is where science ends and metaphysics begins. In our tale, Quine launches ontology into respectability by showing it to be of a piece with science, and thus not sharply differentiated from science (Quine 1951; but see Price 2009 for a different story). More recently we have, for example, Paul (2012) characterizing the difference as one of subject matter, not method, and Morganti (2013) disagreeing. However, the authors in this volume do not consider the question in much depth, even as they discuss significant differences between metaphysics and science. Only Ladyman even addresses the question, and he characterizes the difference merely institutionally (following Ladyman et al. 2007). It is worth pausing to consider why this question has not captivated our assemblage of philosophers of science. Let me give a speculative answer, in two parts.

One part of the answer is that philosophers of science, more than others, are children of Quine, and consider the question settled: there is no substantive, sharp distinction between science and metaphysics (and other parts of philosophy), just as Quine said. After all, philosophers of science are specifically drawn to questions that involve detailed engagement with science, and are in the habit of thinking across the divide.

A second answer, compatible with the first, is that philosophers of science have learned not to worry about drawing such lines because of the history of the demarcation problem. Coming up with criteria to distinguish science from nonsense, or pseudo-science, or non-science was an important task for philosophers of science throughout much of the twentieth century. By the end of a century in which no consensus was reached, although the question remained of great interest, few remained committed to developing precise criteria to characterize science and distinguish it from other things. Indeed, there has been far more consensus that things like astrology and creationism are not science than there has been on any general criteria, and philosophers of science have been more focused on arguments against such specific claims to scientific status (as they have focused more on specifics in general). So perhaps philosophers of science, happy to work with a conception of science that is not cleanly delineated from other forms of inquiry, have thus not seen the need to think hard about the distinction between metaphysics and science *in general*. Rather, as with pseudo-science, they recognize metaphysics when they see it, and that is good enough.

It is now common to draw a distinction between metaphysics and metametaphysics (although, just as a language may contain its own metalanguage, it seems that metaphysics contains metametaphysics). Similarly, we might draw a distinction between discussions that engage in metaphysics and philosophy of science and discussions that are about metaphysics and philosophy of science. This distinction provides a rough two-part structure to the volume. The first part contains essays that primarily engage with metaphysical questions that arise in the context of philosophy of science: philosophy of physics (Brading and Ladyman); philosophy of biology (Slater and Waters); philosophy of ecology (Strevens); and interpreting scientific statements of law and probability (Ismael). The essays in the second part are more methodologically oriented, and consider such questions as what kind of metaphysical philosophy of science is possible (Stanford); how the methods of science relate to the methods of metaphysics (Saatsi); and to what extent philosophy of science is independent of metaphysics (Woodward and Thomson-Jones). This division into two parts is a bit rough, as the more "ground-level" essays in the first part all draw methodological lessons from the specific questions they tackle, and the essays in the second part all illustrate their discussions by appealing to details of science and philosophy of science. Indeed, I am not sure Ladyman's contribution can be very easily assigned to only one category. Nonetheless, the division provides one more useful conceptual distinction for thinking through the relation between metaphysics and the philosophy of science.

It is not for philosophy of science to render a verdict on metaphysics. All previous efforts to make philosophy "scientific" were not able to kill metaphysics, and doubtless the contemporary resurgence of criticism will not either. But metaphysics in the twentieth century certainly evolved under the pressure of developments in science and philosophy of science, and we should expect further evolution in this century as well. We hope that the essays in this volume are able to play a role in that evolution.

REFERENCES

Ayer, A. J. 1936. *Language, Truth and Logic*. New York: Penguin.

Balashov, Yuri, and Alexander Rosenberg, eds. 2002. *Philosophy of Science: Contemporary Readings*. New York: Routledge.

Callendar, Craig. 2011. "Philosophy of Science and Metaphysics." In *The Continuum Companion to the Philosophy of Science*, edited by Steven French and Juha Saatsi, 33–54. New York: Continuum.

Chakravartty, Anjan. 1998. "Semirealism." *Studies in History and Philosophy of Science* 29: 391–408.

Chakravartty, Anjan. 2011. "Scientific Realism." In *Stanford Encyclopedia of Philosophy*. Stanford University, 1997–. Article published April 27. http://plato.stanford.edu/archives/spr2014/entries/scientific-realism/.

Chalmers, David, David Manley, and Ryan Wasserman, eds. 2009. *Metametaphysics: New Essays on the Foundations of Ontology*. Oxford: Oxford University Press.

Creath, Richard. 2011. "Logical Empiricism." In *Stanford Encyclopedia of Philosophy*. Stanford University, 1997–. Article published April 4, revised September 19. http://plato.stanford.edu/archives/spr2014/entries/logical-empiricism/.

Curd, Martin, J. A. Cover, and Christopher Pincock. 2012. *Philosophy of Science: The Central Issues*. New York: Norton.

Dennett, Daniel. 1991. "Real Patterns." *The Journal of Philosophy* 88 (1): 27–51.

Dorr, Cian. 2010. Review of *Every Thing Must Go: Metaphysics Naturalized*, by James Ladyman, Don Ross, David Spurrett, and John Collier. *Notre Dame Philosophical Reviews*. http://ndpr.nd.edu/news/24377-every-thing-must-go-metaphysics-naturalized/.

Kuhn, Thomas. (1962) 1996. *The Structure of Scientific Revolutions*. Chicago: University of Chicago Press.

Hacking, Ian. 1983. *Representing and Intervening*. Cambridge: Cambridge University Press.

Ladyman, James, Don Ross, David Spurrett, and John Collier. 2007. *Every Thing Must Go: Metaphysics Naturalized*. New York: Oxford University Press.

Lowe, E. J. 2011. "The Rationality of Metaphysics." *Synthese* 178: 99–109.

Morganti, M. 2013. *Combining Science and Metaphysics: Contemporary Physics, Conceptual Revision and Common Sense*. London: Palgrave Macmillan.

Paul, L. A. 2012. "Metaphysics as Modeling: The Handmaiden's Tale." *Philosophical Studies* 160: 1–29.

Price, Huw. 2009. "Metaphysics after Carnap: The Ghost Who Walks?" In *Metametaphysics: New Essays on the Foundations of Ontology*, edited by David Chalmers, Ryan Wasserman, and David Manley, 320–46. New York: Oxford University Press.

Psillos, Stathis. 1999. *Scientific Realism: How science tracks truth*. New York: Routledge.

Ross, Don, James Ladyman, and Harold Kincaid. 2013. *Scientific Metaphysics*. Oxford: Oxford University Press.

Quine, Willard Van Orman. 1948. "On What There Is." *Review of Metaphysics* 2 (5): 21–36.

Quine, Willard Van Orman. 1951. "Two Dogmas of Empiricism." *Philosophical Review* 60 (1): 20–43.

Quine, Willard Van Orman. 1960. *Word and Object*. Cambridge, MA: MIT Press.

Sider, Theodore. 2008. "Introduction." In *Contemporary Debates in Metaphysics*, edited by Theodore Sider, John Hawthorne, and Dean W. Zimmerman, 1–7. Malden, MA: Blackwell.

Sider, Theodore, John Hawthorne, and Dean W. Zimmerman, eds. 2008. *Contemporary Debates in Metaphysics*. Malden, MA: Blackwell.

Van Fraassen, Bas C. 1980. *The Scientific Image*. New York: Oxford University Press.

Van Fraassen, Bas C. 1989. *Laws and Symmetry*. New York: Oxford University Press.

Van Inwagen, Peter. 2007. "Metaphysics." In *Stanford Encyclopedia of Philosophy*. Stanford University, 1997–. Article published September 10. http://plato.stanford.edu/archives/win2013/entries/metaphysics/.

Worrall, John. 1989. "Structural Realism: The Best of Both Worlds?" *Dialectica* 43: 99–124.

1

Time for Empiricist Metaphysics

Katherine Brading

I. INTRODUCTION

To what extent are the details of empirical enquiry relevant for the metaphysics of time? I shall argue that they are deeply, utterly, and inextricably entwined, and moreover that they became so as a consequence of philosophical moves made by Newton in his *Principia*.[1] Prior to the *Principia*, general questions about the nature and structure of time, such as whether or not time is merely an aspect of material change, whether there is one time or many, whether time is inherently metrical, and so forth, could be (and were) appropriately addressed via arguments based on broadly a priori considerations. In the wake of Newton's *Principia*, this is no longer the case. Newton showed how the answers to these questions depend on the intricate details of empirical enquiry. Those of us who are interested in the metaphysics of time are not free to pretend that the philosophical moves made by Newton were never, in fact, made. He made these moves, and this paper is about their implications for the metaphysics of time.

In the recent *Blackwell Companion to Philosophy of Time* (2013), there is a section on "The History of the Philosophy of Time," and there, between a chapter on creation and eternity in medieval philosophy and one on classical empiricist discussions of time, we find a chapter on Newton. In this chapter, Eric Schliesser asks us to pause with Newton, and to look in more detail at his contributions to

[1] References are to Newton (1999).

the philosophy of time. Instead of taking Newton's physics, and then looking at what other philosophers have to say about time in the light of his physics, we look at what philosophical moves Newton himself made. I think Schliesser is right that this is worth doing, and I claim that Newton's empirical methods reach deeply into metaphysical questions concerning the nature and structure of time. For philosophers with an interest in the metaphysics of time, Newton's *Principia* needs to be read as a philosophical text, offering contributions to an empiricist metaphysics of time.[2]

I begin from three distinctions that Newton made at the beginning of his *Principia*, in the famous scholium on time, space, place, and motion. There, he said that we should distinguish between absolute and relative, true and apparent, and mathematical and common, for each of time, space, place, and motion. I outline these distinctions as they apply to time (section 2), and then discuss Schliesser's (2013) interpretation of Newton's distinction between absolute and true time (section 3), explaining why I think a different approach is needed. I then build toward the positive conclusions that I want to draw. I begin by offering an alternative interpretation (section 4), according to which Newton is drawing on existing terminology and implicit conceptual distinctions in order to make explicit and systematic a three-way set of distinctions concerning the nature and structure of time. In so doing, he makes a contribution to the philosophy of time. I then argue that (a) these distinctions are empirically accessible (see section 5), and (b) all three distinctions are necessary for setting up the project of the *Principia* (see section 6). It follows from this, I argue, that certain questions concerning the nature and structure of time become empirically tractable through the pursuit of that project, or some appropriately similar project. By situating Newton's *Principia* in the appropriate philosophical context (section 7), we can read the *Principia* as a direct contribution to the metaphysics of space, time, matter, and motion, and as offering an empiricist metaphysics of a particular kind. I claim that Newton refined the conceptual distinctions appropriate for asking questions about the nature and structure of time, and transformed the methodology by which such questions should be addressed, and I show this in detail for each of the three distinctions he makes at the outset of the *Principia* (see section 8). I conclude by drawing together the results of the preceding sections, and then use the specific example tackled in this paper to suggest some more general lessons about how philosophers should approach the relationship between metaphysics and empirical science (section 9).

---

[2] I take the phrase "empiricist metaphysics" from Janiak (2008, 29) who, in describing the re-appraisal of Newton due to Stein, writes, "The result is not an anti-metaphysical Newton but a kind of empiricist metaphysician." See also Stein (1970).

2. THREE DISTINCTIONS

Newton's *Principia* (published in 1687) opens with a series of definitions of the terms that he will use, including "quantity of matter," "quantity of motion," "inherent force of matter," and so forth. Immediately following these definitions he turns his attention to time, space, place, and motion, in a scholium that begins as follows (Newton 1999, 408):

> Although time, space, place and motion are very familiar to everyone, it must be noted that these quantities are popularly conceived solely with reference to the objects of sense perception. And this is the source of certain preconceptions; to eliminate them it is useful to distinguish these quantities into absolute and relative, true and apparent, mathematical and common.

The literature has largely focused on absolute versus relative motion, and absolute space, with comparatively little discussion of time,[3] and nothing that I know of about why Newton has this *three-fold* set of distinctions.

However, Schliesser's (2013) discussion of Newton on time was inspired by Huggett (2012), who offered an interpretation of the distinction between absolute and true *motion*.[4] I differ from both in my interpretation of the terminology, and therefore in the distinctions that Newton is drawing. In this paper I focus on the case of time.

Newton introduces his discussion of time as follows:

> Absolute, true, and mathematical time, in and of itself and of its own nature, without reference to anything external, flows uniformly and by another name is called duration. Relative, apparent, and common time is any sensible and external measure (precise or imprecise) of duration by means of motion; such a measure—for example, an hour, a day, a month, a year—is commonly used instead of true time. (Newton 1999, 408)

Thus, Newton is explicit in applying to time the three distinctions of absolute versus relative, true versus apparent, and mathematical versus common. What does he mean by these distinctions, and why do they matter?

---

3 Arthur 1995; Gorham 2012; McGuire 1978; Palmerino 2013.

4 The disagreements that I have with Schliesser and Huggett are small compared to the overall content of their papers, from which I learned much. As so often happens, the many points of agreement and enlightenment I pass over in silence, to focus on what we may learn from a point on which we disagree.

According to Schliesser (2013), the terminology of "absolute" and "true" marks a distinction between time as a theoretical construct, arrived at from empirical considerations and applied in the context of the project of the *Principia* (absolute time), versus time as it occurs in metaphysical and theological discussions (true time). He writes,

> True time is an unnecessary addition to Newton's conceptual framework of absolute and mathematical time given the particular problems addressed in the *Principia*. (91)

And he goes on,

> Newton introduces more conceptual distinctions than are required by his physical theory; his dynamics requires no more than absolute (mathematical) time as a contrast to "relative, apparent, and common time" without resort to "true" time. (92)

On Schliesser's view, while it plays an important role in Newton's overall philosophy of time, "true" time (and therefore the contrast between true and apparent time) has no place in the empirical project of the *Principia*.

I disagree. And what is at stake here is not merely Newton exegesis: resolving the terminological issue has implications for which questions concerning the nature and structure of time Newton's empirical project is capable of addressing. In other words, what is at stake is the metaphysical reach of his empirical methods. I think, *contra* Schliesser, that *all three* of Newton's conceptual distinctions between absolute and relative, true and apparent, and mathematical and common time, have empirical import (section 5) and are necessary for the project of the *Principia* (section 6). Moreover, I maintain that these distinctions bear on long-standing philosophical questions concerning the nature and structure of time (sections 7 and 8). As a consequence, these questions become subject to empirical investigation in a manner not possible prior to the relevant philosophical innovations of Newton's *Principia*.

In the following section I discuss Schliesser's approach in some more detail, before turning to my preferred approach in section 4, below.

3. Schliesser's Account of Newton on Time

Schliesser (2013) argues that we should distinguish absolute time from true time, and I think he is right that the two terms are not synonymous for Newton. However, whereas for Schliesser "absolute time" is a mathematical construct, a time parameter in our physical theory by which we seek to place a measure on the passage of true time, I will argue below that "absolute" and "true" are both to be understood as characteristics of time, characteristics that will both be represented in an appropriate time parameter of an adequate physical theory.

Schliesser's proposal is that absolute time is a theoretical construct obtained from our observations of the heavenly bodies and their relative motions via the equation of time (of which more below). He writes,

> As a first approximation, we can say that "absolute" time is approximated by our clocks (or some other measure of relative time) corrected by the astronomical equation of time . . . allowing thus a *measure* of true time. (2013, 90)

On this view, the empirical basis of "absolute time" is such that it extends spatially only so far as the bodies in our planetary systems whose relative motions we observe (with respect to the background of the fixed stars). True time, by contrast, is that which we are seeking to approximate in our construction of absolute time, and, according to Schliesser, it extends from infinity to infinity. Moreover, insofar as there is always room for improvement in the process by which we construct absolute time, we can think of it as a "regulative ideal."

Schliesser's "absolute time" is needed for the project of the *Principia* because what Newton is trying to do in the *Principia* is to determine whether our planetary system is geocentric or heliocentric, and in order to solve this problem he needs to be able to assign accelerations to bodies in our planetary system in a unique and consistent way. This latter task requires a single time parameter relative to which the bodies in our planetary system are assigned their motions. Furthermore, for the purposes of the project, the spatial extent of this time parameter need be only so far as the observed bodies in our planetary system, and this is consistent with the empirical basis from which the time parameter is constructed.

For Schliesser, our empirically constructed (absolute) time is only approximately accurate compared to metaphysical (true) time, and it has limited spatial extent. When Newton introduces true time, in addition to absolute time, he does not thereby introduce any new empirical questions, because true time is answerable to the demands of Newton's rational theology and not to empirical considerations. The only open empirical question concerns the spatial extent of absolute time: will the time parameter appropriate for our planetary system turn out to be appropriate for treating bodies *outside* our planetary system?

This is one way to read the distinctions, and to think about their significance for philosophy of time. I agree up to a point. I agree that Newtonian absolute time should not be conflated with Newtonian true time. I agree that, for the purposes of the *Principia*, Newton does not need his time parameter to extend from infinity to infinity. I agree that the spatial reach of Newton's time parameter in the *Principia* is an empirical matter. However, I do not think that, for Newton, "absolute" and "true" mark Schliesser's distinction between a spatially limited empirical time parameter and a theologically motivated, infinitely extended "time."

Initial doubts about Schliesser's interpretation arise when we notice that in the scholium Newton does *not* make the positive assertion that absolute, true, or mathematical time are eternal in duration, nor does he assert that space is infinite, and nor does he assert that each moment of time extends from infinity to infinity. We are familiar with these claims from other places in Newton's writings, but in this part of the text, where Newton is setting out what is needed for the project of the *Principia*, no such positive claims are made.

Moreover, the distinction that Schliesser draws is not one that we find doing work for Newton in his argument in the *Principia*, such that he has reason to mark it by means of a terminological distinction. As evidence for this, consider that Newton has just as good reasons to think that his absolute time extends to the physics of the distant stars and to the planetary systems around distant stars (if any such exist) as he does to think that his laws of motion and law of universal gravitation apply to such bodies, and he does not make the solar system the boundary of applicability for these latter. Newton worried about how we extend our knowledge to bodies beyond the reach of our experiments, and this worry is explicitly addressed in his Rule 3 of Reasoning, added to Book 3 in the second edition of the *Principia* (Newton 1999, 795):

> Those qualities of bodies that cannot be intended and remitted and that belong to all bodies on which experiments can be made should be taken as qualities of all bodies universally.

This rule plays a crucial role in enabling Newton to extend results from terrestrial experiments to the celestial bodies of the solar system. In applying Rule 3 to bodies beyond the solar system, we would certainly be wise to be tentative given the flimsiness (even non-existence) of our empirical evidence, but there is nothing in Newton's writings to indicate a sharp cut-off at the outer edges of the solar system such that we should not consider distant stars to be bodies. On the contrary, the possibility of other worlds around other Suns, governed by the same laws, is very much part of Newton's thinking. For example, there is a manuscript[5] in which Newton

[5] See Hall and Hall 1962, 374–77.

asserts that the fixed stars are bodies just like our Sun: they are formed into spheres by their own gravity, and since they are bodies, they are, by definition, subject to the laws of motion. It seems to me that the distinction Schliesser draws is not important for Newton's purposes.

Finally, the contexts in which Newton extends moments of time to spatial infinity are generally also those in which he is talking about God's presence in the world, rather than those in which he is concerned with methods of reasoning from the phenomena.

In my opinion, we have good reason to suspect that Newton was careful not to overreach empirically when he was setting out his accounts of time and space in the scholium (i.e., at the outset of the empirical project of the *Principia*). There is therefore reason to doubt that the inclusion of "true time" is an "unnecessary addition" (Schliesser 2013, 91). In the following section I propose an alternative interpretation of the terminology, in which each of his three distinctions—absolute versus relative, true versus apparent, and mathematical versus common—are relevant and important for the project of the *Principia*.

4. AN ALTERNATIVE INTERPRETATION

When it comes to motion, there is a standard way of understanding the absolute versus relative terminology: relative motion is motion of a body with respect to other bodies, and absolute motion is motion of a body independent of its relations to other bodies (but perhaps in relation to absolute space and time instead). I take this to be uncontroversial. The idea of motion as relative to other bodies was familiar (from Descartes, among others), and much discussed, and moreover this is the one distinction that Newton himself explains in detail. He uses it in the same way for space, time, and place as he does for motion, and absolute motion is further spelled out in terms of absolute space and absolute time. Newton is clear about this terminology, and I take it that his readers would have understood his use of the terms in the way that is now standard. Absolute time is independent of material bodies, whereas relative time is an aspect of material bodies or of the relations among them.

The terms "true" and "apparent" have meanings that would already have been familiar to Newton's readers from the dispute over the Copernican system, so some historical context is useful. The Copernican dispute concerns whether the system of the world is geocentric or heliocentric or whether there is no fact of the matter. Book III of Newton's *Principia* is called "The System of the World," and this is where Newton marshals the resources developed in Books I and II to give his answer to the Copernican question. Addressing this question is the overall purpose of the *Principia*, and the Copernican dispute is therefore the appropriate context for understanding what Newton means by the terminology "true and apparent" in the *Principia*.

Within this dispute, those in the geocentric and heliocentric camps shared a commitment to true motion, as distinct from apparent motion: whatever the apparent motion of a given body (e.g., the Sun moving across the sky), and of which there may be many (depending on the position of the observer), there is nevertheless a unique motion that is the true motion of that body. Thus, for geocentrists and heliocentrists alike, one motion is singled out as not mere appearance, but proper to the body, and this is its true motion. The distinction between true and apparent motion comes to the fore in the Copernican dispute because of the obvious conflict in Copernicus's system between the appearances (it appears to us, observing from Earth, that the Sun moves around a stationary Earth) and the true motions described in the theoretical system (the Earth in motion around a stationary Sun), but even those astronomers committed to a geocentric theory made the distinction between the apparent motions of the heavenly bodies (how their motions appear to us) and their true motions.

With these distinctions in hand (between absolute versus relative and true versus apparent) there remains the issue of the relationship between them. For example, Descartes sought to give a relative account of true motion, whereas Newton sought to define true motion in terms of absolute motion. In his *Principles* II.25, Descartes (1991) defined true motion in terms of the relative motion of a body with respect to the immediately surrounding bodies (themselves considered to be at rest), and thereby sought to identify the one true relative motion proper to the body (i.e., its true motion) from the many relative motions that this body in the plenum undergoes. One of the things that Newton tried to do in the scholium to the definitions in the *Principia* was to show that true motion should be identified not with relative motion but with absolute motion, where for Newton absolute motion is motion with respect to absolute space and absolute time.

So these distinctions, and also I think the terminology, were already out there for the case of motion. Newton gathers them together, sets them out as two distinct pairs, and standardizes them across space, time, place, and motion. If this is the right way to approach the terminology,[6] then the contrast between true and apparent time is the following. Just as true motion is unique and proper to the body (or system of bodies) in question, in contrast to being a property of the appearances, so too true time is unique and proper to the body (or system of bodies) in question, a property of the body or system itself rather than of the appearances. We can illustrate the idea as follows. The relative motions of the Sun and Moon with respect

[6] See Huggett (2012) for an alternative.

to the Earth give rise to their apparent motions with respect to the fixed stars, as viewed from Earth. Each apparent motion may be used as a clock for the Earth-Sun-Moon system, and therefore each gives rise to an apparent time (apparent solar time and apparent lunar time respectively). But these "clocks" do not tick regularly with respect to one another, and therefore the true time of the Earth-Sun-Moon system (if any such exists) remains to be determined.

The terminology of "mathematical" and "common" has its origins not in discussions of motion, but in the treatment of time in mathematical astronomy. In the scholium, Newton writes that absolute time is distinguished from relative time by the equation of common time. We can therefore begin our investigation of what is meant by "common time," and by the contrasting term "mathematical time," by looking at the equation of common time. Common time, in this context, is time on which a metric is imposed by means of material clocks, for the purposes of our common life; it is the division of the passage of time into intervals by which the rhythm of our lives is marked out, and has no more precision, nor any other properties, than those needed for this purpose.[7] In the seventeenth century, apparent solar time (of which more below, see section 5) was used as the basis of common time. By contrast, the equation of common time was used by astronomers to construct a time parameter suitable for the purposes of mathematical astronomy (as explained below, section 5). This time parameter has precise mathematical properties (it is metrical, it is continuous, and so forth), and in later sections of this paper I will stress the importance of Newton's time parameter being metrical. This time parameter was of no interest or use to anyone except mathematical astronomers. For Newton's intended readership, well versed in the problems of horology and mathematical astronomy (especially Huygens), this terminology of mathematical and common time would have been readily understood.

I claim, therefore, that all six terms in the three distinctions have meanings that are prior to, and external to, the project of the *Principia*. What is new with Newton is stating them all explicitly as contrasts like this, stating them all together, and applying them uniformly and systematically across time, space, place, and motion.

If this is right, then all three distinctions are associated with independent questions concerning the nature and structure of time. Is time absolute or relative? Is time true or apparent? Is time mathematical or common? Moreover, as I will argue in what follows, each distinction has empirical import (section 5), all three conceptual distinctions are needed for setting up the project of the *Principia* (section 6), and therefore

[7] It would be interesting to know whether there is any connection between this use of the term "common time," and that in music (current at the time), where 4/4 time was considered "imperfect" and was known as "common time," whereas time signatures with a three-measure (e.g., 3/4 time) were considered "perfect" (in accordance with the Trinity).

each of the three questions becomes subject to empirical investigation (see also section 6). The upshot is that, at the very least, there are more open empirical questions concerning time than Schliesser's interpretation of this terminology would allow. But there is much more than this. In making these distinctions explicit for the first time, and in tying them to the details of empirical enquiry, Newton makes the questions about the nature and structure of time more fine-grained, and transforms the process by which we are to address them. I will argue (section 8) that all three distinctions engage with familiar metaphysical questions concerning the nature and structure of time, and so those questions themselves become empirically tractable in Newton's hands. His work in the *Principia* thus constitutes an important transformation in the appropriate methodology for pursuing the metaphysics of time.

5. EMPIRICAL ACCESS

With the distinctions between absolute and relative, true and apparent, and mathematical and common understood as I suggest, I will now argue that they are empirically accessible, and therefore subject to empirical investigation. Moreover, this is something of which Newton was well aware at the time he was writing the *Principia*.

Our empirical access to time is indirect, via material bodies: we make use of material systems that exhibit periodic phenomena, that is, which tick. The most obvious periodic phenomenon in our daily lives is the cycle of day and night itself. Traditionally, and into the fifteenth century, the hour was defined by taking the time from sunrise to sunset and dividing it into twelve (and similarly for the twelve hours of the night).[8] Call this seasonal time. Time understood in this way is relative (it depends on the relative motion of the Sun and the Earth) and apparent (deriving from the motion of the Sun as it appears to us).

How would the division of day and night each into twelve hours be achieved? During the day, the Sun appears to us to move steadily across the sky, so the apparent motion of the Sun (assumed to be constant) could be used to divide the day into twelve. During the night, the stars similarly appear to move steadily across the sky, so the apparent motion of the stars (assumed to be constant) could be used to divide the night into twelve. One upshot of this approach seems to be that daytime hours in summer are longer than daytime hours in winter. But how can that be? We defined the hour as our unit of time, and by definition each hour must be of the same duration: an hour. So what does it mean to say that hours vary in duration,

[8] For a detailed discussion of the historical and conceptual development of time measurement, see Audoin and Guinot 2001.

and how could we show that this is the case? One way to tell is to compare these seasonal hours with other periodic phenomena, such as the number of times sand runs through an old-fashioned egg-timer (or an hourglass!) in the course of an "hour"; even our own biological rhythms are sufficient for us to be able to "experience" a difference in length of seasonal hours.

In other words, there are different apparent times, arising from different choices of bodies whose relative motions we use to construct a clock. Faced with such disagreement, it can be convenient for a society to make a choice that is shared among its members. We adopt a common time, and in the mid-seventeenth century, apparent solar time was used for this purpose. Apparent solar time is defined as follows: take one complete circuit of the Sun around the Earth and divide it equally into 24 hours, so that one hour corresponds to the Sun moving through an angle of 15°.

Apparent solar time is a relative conception of time (depending on the relative motion of the Sun and the Earth). During the day, it is directly observable (look and see how long it takes for the Sun to move through 15°, or measure this indirectly by means of a sundial). During the night, we need theory to calculate how far the stars have to move for a corresponding 15° motion of the Sun (postulated to be continuing its motion around the other side of the Earth). With this calculation in hand, such "apparent solar time" is observable—it is in the appearances and is therefore apparent.

When compared to other periodic phenomena, this is a great improvement: those phenomena are now in much greater agreement with our chosen clock. So there is good reason for ordinary people to switch from seasonal time to apparent solar time as the basis for common time, and many cities and towns did so. According to Audoin and Guinot (2001, 40), apparent solar time was in use as the standard for common time in country areas of Europe until the beginning of the twentieth century.

Astronomers too would have had good reason to prefer apparent solar time over seasonal time as their time parameter for astronomy. They were engaged in the task of predicting the motions of the heavenly bodies (the stars, the Sun, the Moon, and the planets) using uniform circular motion. Any deviations of these bodies from uniform circular motion needed to be accounted for theoretically. By adopting apparent solar time rather than seasonal time, the motions of the heavenly bodies become less irregular. That is to say, they are less irregular with respect to equal intervals of time defined using apparent solar time. The role of theory in our understanding of time is clear: if one choice of clock (the standard of time that we adopt) yields the result that the motions are highly irregular, while another choice of clock yields the result that the motions are close to regular, then the second clock makes our theoretical task easier as we try to account for remaining irregularities.

Despite its great practical utility as a basis for *common* time, apparent solar time was nevertheless deemed "irregular" even by ancient astronomers. They worked instead with mean solar time, which is constructed as follows. Every day, when the Sun rises, it rises at a slightly different point relative to the background of the stars. A solar year is the time taken for the Sun to rise again at that same point plotted against the background of the stars. In the course of this year, if we plot the position of the Sun at sunrise every day with respect to the stars, we see that the Sun makes a complete journey around the sky, along a path called the ecliptic. This path is so important that we keep track of it at night (when we cannot see the Sun) by means of the constellations of the zodiac. But the Sun speeds up and slows down during the year (passing through a greater or lesser angular distance along the ecliptic on different days). If we smooth out the motion of the Sun, so that it moves at a constant daily speed around the ecliptic, the resulting position with respect to the fixed stars is the position of the mean Sun.[9] Except when the actual Sun and the mean Sun coincide, no material body is located at the position of the mean Sun. But, if we use the mean Sun as our clock, the motions of the heavenly bodies overall become more regular: regularities in the irregularities are removed. This serves our goal of constructing a predictively adequate theory using regular motions: irregularities in the appearances (the apparent motions) are to be accounted for by such things as the eccentricity of the Sun's orbit, and so forth. Mean solar time "corrects" apparent solar time by removing periodic irregularities via the "equation of common time."[10] Notice that we are using theory here to move from apparent time, derived from the most regular apparent motions that we observe (celestial motions), to an abstract theoretical time. Mean solar time is a theoretical construct: no material bodies used in its construction were observed to move regularly with respect to mean solar time, and it was therefore neither apparent nor relative. *Astronomers knew of no relative motions that could serve as a clock with respect to this time.* Thus, the "time" that is appropriate for astronomy is mathematical, and it is neither relative nor apparent.

---

[9] Notice that this attributes all the irregularities to the Sun. The motion of the stars with respect to the Earth is assumed to be uniform, and this is used as the regular background with respect to which the apparent motion of the Sun is then smoothed out to construct the mean solar time. The uniform motion of the stars is equivalent to assuming that the daily rotation of the Earth is uniform. According to Audoin and Guinot (2001, 46–48), Kepler mentioned the possibility of some irregularities in the Earth's rotation, Flamsteed was the first (1677) to try to detect them (without success), Maupertuis (1752) wondered whether there might be some irregularities and, if so, what their cause could be, and Kant suggested that "there could be slowing down effect due to dissipation of energy in tidal movements of the oceans. He was right, but the idea was only confirmed by observation two centuries later." Laplace (1825) rejected the possibility out of hand. It was not until the twentieth century that tiny irregularities in the rotation of the Earth were confirmed. See Smith 2014, 302–07.

[10] For more details, see Audoin and Guinot 2001, 40ff.

This is a conclusion that we arrive at through an interplay between theoretical and empirical considerations, in which a theory of planetary motions is constructed by assigning simple basic motions to those planets and then treating the deviations as corrections to those basic motions. The concepts of relative, absolute, apparent, true, common, and mathematical time were all present in the development of the project of constructing a time parameter appropriate for the purposes of mathematical astronomy, as that project existed at the time that Newton was writing, and therefore all three distinctions were empirically engaged at that time. These are distinctions that we can get at empirically, through the project of mathematical astronomy.

The intricacies concerning the treatment of time in mathematical astronomy were familiar to Newton. In the scholium to the definitions he writes,

> In astronomy, absolute time is distinguished from relative time by the equation of common time. (Newton 1999, 410)

In other words, we move from common time (a time adequate for the purposes of regulating our daily lives) to a time parameter appropriate to the needs of mathematical astronomy, and in so doing we move from relative time, on which common time is based, to absolute time, making use of the equation of common time and therefore of the mathematical properties of our time parameter. Newton further elaborates on the role of the equation of common time as follows:

> [D]uration is rightly distinguished from its sensible measures and is gathered from them by means of an astronomical equation. Moreover, the need for using this equation in determining when phenomena occur is proved by experience with a pendulum clock and also by eclipses of the satellites of Jupiter. (Newton 1999, 410)

The point he makes in this paragraph is that, in practice, we have strong theoretical reasons for believing that we have not yet found bodies (either celestial or terrestrial) whose periodic motions can serve as perfect clocks. The time parameter that would be measured by such perfect clocks, and that we have strong empirical reasons for adopting based on our investigations in astronomy and in terrestrial clock making, is neither relative nor apparent. Thus, Newton knew very well that the "time" that is appropriate for astronomy, and thus the "time" that is appropriate for solving the problem of the system of the world, is mathematical, and is neither relative nor apparent. In specifying his three distinctions, he knew that they are distinctions that we can get our hands on empirically, and that we do this via the theories that we construct in order to account for the detailed motions of celestial bodies (the planets) and of terrestrial bodies (pendulum clocks).[11]

6. ABSOLUTE, TRUE, AND MATHEMATICAL TIME IN THE PROJECT OF THE *PRINCIPIA*

We have seen that Newton had some empirical basis for making the distinctions between absolute and relative, true and apparent, and mathematical and common time (as I have interpreted that terminology), and that he had good empirical reasons for believing that the time parameter appropriate for mathematical astronomy is absolute, true, and mathematical, rather than relative, apparent, and common. I will now argue that each of these three conceptual distinctions suggested by mathematical astronomy is *necessary* for setting up the project of the *Principia*. This enables me to argue that the project of the *Principia* is a means for empirically investigating the characteristics of time that are associated with its being absolute or relative, true or apparent, and mathematical or common.

As noted above, the end goal of the *Principia* is to establish the "system of the world"—that is, to decide between the traditional geocentric, the Tychonic geocentric, and the heliocentric systems by establishing the true motions of the bodies in our planetary system.[12] Successful completion of the project of the *Principia* thus requires true motion: without true motion, there is no determinate answer to the question of the system of the world. The project therefore requires the distinction between true and apparent motion.

Newton's proposed strategy for solving the problem of the system of the world was to systematically correlate forces with true motions. In the scholium to the definitions, Newton argues that true motion cannot be relative motion, and must therefore be absolute motion instead, by showing that on a relative conception of motion, true motions are not appropriately correlated with the presence/absence of

[11] I have argued that Newton had good reason to believe that no material system is a perfect clock, and that time should therefore be considered absolute and not relative. My focus was on astronomical clocks, but Schliesser (2013) reminds us of the importance of seventeenth-century advances in pendulum clocks and their connections with timekeeping in astronomy. Newton was, of course, deeply immersed in this work too, especially through the tight interconnections between the study of the pendulum and of gravitation, and he was intimately engaged in studying the precise limitations and approximations involved in pendulum clocks. This work serves to reinforce the need for a distinction between the material clock and the absolute time that it approximates. See Schliesser's paper for much greater contextualization of Newton's treatment of time than I have given here.

[12] In the end, of course, Newton replaces all of these with a system in which none of these bodies remains at rest at the center of the system. Nevertheless, in doing so he establishes the true motions of these bodies.

forces. This is one purpose of the famous bucket experiment.[13] Thus, insofar as true motion is necessary for the project, we also need absolute motion. But we cannot have absolute motion with relative time, since the resulting motion would then be dependent on the material bodies whose relative motions constitute relative time. Therefore, successful completion of the project of the *Principia* requires the distinction between absolute and relative time.

True motion also requires true time: there must be a unique time parameter proper to the system of the world, for if there is not, then a second "time" might give different conclusions concerning the motions and forces, and therefore concerning the system of the world. Thus, setting up the project of the *Principia* requires that we distinguish between true and apparent time.

If mathematical astronomy is the appropriate route for solving the system of the world, as Newton believed, then the need for distinguishing between mathematical time (with its metrical properties, and so forth) and common time is demonstrated by the need for an equation of common time. As noted above, Newton's proposed strategy for solving the problem of the system of the world was to systematically correlate forces with true motions, which in turn demanded a mathematical treatment of forces and motions as set out in Books I and II of the *Principia*. The specific mathematical properties required of the time parameter in these treatments stands in need for further investigation by Newton scholars,[14] but it is immediately clear from the outset of the *Principia* that the time parameter must be metrical: Law 1 relies on equal intervals of time for the distinction between uniform and non-uniform motion, and this is at the heart of the distinction between the presence and absence of forces by which we are to arrive at the true motions.

The distinctions between absolute and relative, true and apparent, and mathematical and common time are therefore intimately tied to the project of the *Principia*. Newton had good reason to assert all three distinctions in setting up his project. The three distinctions are mutually independent of one another, and all three are needed for the purposes of his empirical project.

Newton also had good reason to believe from the outset that the time parameter appropriate for successful execution of his project would turn out to be absolute, true, and mathematical, rather than relative, apparent, or common. However, setting Newton's own position to one side, one cannot know at the outset whether the demands of the empirical project, as it unfolds, will indeed restrict the characteristics of the time parameter such that it turns out to be absolute, true, and

---

[13] Cohen and Whitman 1999, 412–13. For discussion of the bucket experiment, see Huggett (2012) and references therein.

[14] Arthur 1995; Palmerino 2013.

mathematical. This is something that can be settled only by pursuit of the project. I therefore disagree with Schliesser (2013) that only absolute (mathematical) time is connected to an empirically open question capable of being addressed by the methodology of the *Principia*. As we have seen, Schliesser claims that absolute time (in the sense of a time parameter for the solar system) is needed by Newton for his dynamics in order for Newton to "identify and assign accelerations to moving bodies in a consistent fashion," but that Newton's inclusion of something called "true time" (understood as an extension of absolute time from the "local temporal frame" of the solar system to spatial infinity) turns out to be a metaphysical commitment going beyond the demands of the *Principia*, rooted in Newton's rational theology. I think that Schliesser is right about the need for such a time parameter for the project of the *Principia*, and about its empirical status, but I think that each of the three distinctions (as I have interpreted the terminology) represent open empirical questions concerning the characteristics of the time parameter, and questions that are capable of being addressed by the methodology of the *Principia*.

The three distinctions are subject to empirical investigation in the first place through their connection to the problem of true motion. True motion is necessary for successful execution of the project of the *Principia*, but it is a contingent matter whether any such true motions exist: it might turn out that there are no true motions and thus there is no answer to the problem of the system of the world. Moreover, the project of the *Principia* enables us to probe the connections between true motion and time asserted above: for example, one might interpret Galilean relativity as indicating that true time is necessary but not sufficient for true motion, since there we have true time, and absolute (i.e., not relative) motion, but seem not to have true (i.e., unique) motion. This is an example of how the questions we are asking are transformed and become more fine-grained in the process of addressing them through the project of the *Principia*. There is much more to be said here concerning the execution of the project in the details of the *Principia*, in the uses of time that Newton makes in his mathematical arguments, his search for empirical clocks, and the interplay between these and his construction of absolute, true, and mathematical time. It is by understanding these details, and their relationship to the empirical successes and failures of the *Principia* and later developments in physics, that we will find out the extent to which time turns out to be absolute and/or true and/or mathematical. The point that I want to stress here is that these aspects of the nature and structure of time are now tied to the details of empirical enquiry. All three questions of whether time is absolute or relative, true or apparent, and mathematical or common, have become empirically tractable.

7. EMPIRICIST METAPHYSICS

I have argued that in Newton's hands all three questions of whether time is absolute or relative, true or apparent, and mathematical or common, become empirically tractable through a project such as the *Principia*. But what is the significance of this conclusion for the metaphysics of time? After all, it is the demands of Newton's project that lead to the adoption of an absolute, true, and mathematical time parameter: for the purposes of constructing a physics of matter in motion capable of addressing the question of geocentrism versus heliocentrism, it is absolute time rather than relative time (for example) that serves our needs. And it might be thought, therefore, that the results are restricted to the following form: "For the purposes of solving the system of the world, we must treat time *as if* it is absolute, true and mathematical," and that this claim leaves the deeper philosophical questions concerning the nature and structure of time untouched. But this would be a mistake.

To see why, we need to situate Newton's work in its philosophical context. Newton's *Principia* makes contributions to mathematics, mechanics, astronomy, physics, and philosophy, and the text of the *Principia* should therefore be read in many ways. One of these is as a contribution to philosophy. I argue that we are right to incorporate Newton into the history of philosophy of time, not because his physics has implications for questions in philosophy of time that are to be drawn out by philosophers, but for his own direct contributions to philosophy of time through the conceptual distinctions that he makes and the methodology that he employs.

The most helpful philosophical context for these purposes is the work of Descartes. In Part II of his *Principles of Philosophy*, published in 1644, Descartes set out the framework of a project the goal of which is to provide an account of all the rich variety of the world as we experience it in terms of matter in motion, where that matter is characterized by a minimal set of properties (extension in Descartes's case), "motion" is restricted to local motion (i.e., motion from place to place), and the parts of matter move and interact according to specified laws. Since Descartes's laws are both spatial and temporal, space and time also fall under the scope of his project. It is this portion of Descartes's philosophy that Garber (1992) labeled "Descartes's metaphysical physics." It is indeed a "*metaphysical* physics" in the following sense: it is intended to give an account of the nature, structure, and operation of the material world. The concepts it employs, of matter, motion, laws, and so forth, are not intended to be understood "as if" (i.e., serving an instrumental purpose relative to the project in hand), but as being metaphysically significant. To this metaphysical project, Descartes attached his rationalist epistemology of clear and distinct ideas.

Many of the important philosophical moves made by Newton are best understood as taking place within the framework of this project, and as responses to difficulties that Newton found in Descartes's execution of that project. They should therefore be read as contributions to the metaphysics of matter, motion, space, and time, just as much as Descartes's. What makes this project look so different in Newton's hands is not the project itself, but the epistemology and methodology by which Newton sought to carry it out. Newton's methodology presses us to make the most of the actual world we find ourselves in as an epistemic resource in pursuing the metaphysical project set up by Descartes, and this fastidious attention to empirical details is an expression of Newton's empiricist epistemology. We are being offered an *empiricist* metaphysics, but a metaphysics no less. Thus, for the purposes of philosophy, it is important to read Newton's project as a continuation of Descartes's, and to assess it on these grounds.

As an aside, just to be clear: this is not the classical empiricism that gets discussed in the chapter after Newton in the *Blackwell Companion* (2013) mentioned in the Introduction to this paper. In early modern philosophy, the common starting point for both rationalists and empiricists is "ideas," which are cut free of any connection to the world; this generates the epistemic problem of how to connect them back up again, so that we can know anything about the world. Such a starting point is not neutral, of course, and it is not one that Newton accepts. He is an empiricist, in the sense that all our knowledge comes through our experience; but, for Newton, to know the world through our experiences is not to *first* know our "experiences" and to *then* know the world. He is clear that this is a mistake, and he is clear about the methodological implications that follow.

Challenges to the significance of Newton's work for metaphysics should take place within this philosophical context. For example, one could challenge the value of Newton's contribution to metaphysics by rejecting the goal of Descartes's metaphysical physics (of giving an account of all the rich variety of the world as we experience it) as being irrelevant to the metaphysics of matter, motion, space, and time. The level of detail at which Newton attends to the actual world could perhaps make this move tempting, but I think it would be a mistake. Unless we take the actual world as our guide, we risk constructing a metaphysics of matter, motion, space, and time that fails to include the actual world among its possibilities. Newton's methodology requires us to take very seriously the goal of including the actual world among the possibilities we are addressing, and the metaphysical project set up by Descartes is therefore to be carried out with careful attention to empirical details.

One could also challenge the framework of the project (in terms of matter, motion, and laws) or the methodology adopted for carrying it out. For those of an empiricist inclination, seeing the framework and the methodologies in action will be an important part of this assessment. In being allowed to prove themselves (or not) with respect to the goals of the project, they prove their utility (or not) with respect to the metaphysics of matter, motion, space, and time.

If we adopt this approach, then the framework and the method will be judged by their results, and the significance of the project for metaphysics will depend in part on how fruitful the project proves to be with respect to the goal of giving an account of all the rich variety of the world as we experience it. Failure to succeed with respect to the goal is a legitimate route by which to challenge claims about the significance for metaphysics of the conceptual investigations carried out within the project. Notice that this depends on assessing the *detailed empirical success* of the project, and this is characteristic of the empiricist metaphysics that I am describing in this paper.

My claim is that Newton's *Principia* makes direct contributions to the metaphysics of space, time, matter, and motion, and that this becomes visible if we read the *Principia* as a contribution to philosophy, in its philosophical context, by beginning with our metaphysical questions concerning space, time, matter, and motion as they were prior to Newton, and seeing what happens to them through the process of Newton's attempts to carry out Descartes's metaphysical physics project. In particular, Newton's results concerning time are not of the above "as if" restricted form, but are contributions to our investigations into the nature and structure of time, and can be read as offering an empiricist metaphysics. To see how we address general questions about the nature of time by means of such an empiricist metaphysics, we return to Newton's absolute, true, and mathematical time. For each of the three distinctions that Newton makes, there are issues concerning the nature and structure of time that we are being asked to address, and for each of these Newton is providing empirical purchase such that, as we shall see, addressing these questions is no longer a matter of general philosophical reasoning, but depends on the details of empirical enquiry. For those of us who value history of philosophy as a route toward understanding the philosophical questions that we have today, we know how important it is that we tell and re-tell that history; the case of Newton on time is one example of how we also need to tell and re-tell the history of physics, *read as a contribution to philosophy*, in order to re-tell the history of philosophy.

8. TIME FOR EMPIRICIST METAPHYSICS

8.1 Absolute Time

The question of whether time is absolute or relative concerns the relationship between time and change. The general question is whether time is an aspect of material change, or somehow independent of matter. In a world such as Descartes's, where all change is accounted for in terms of local motion, the question is whether time is an aspect of the relative motions of bodies, or somehow distinct from these. Sharpening the question a little, this becomes: Are there any actual relative motions that can serve as clocks for the material goings-on in the world, such that our project (of providing an account of all the rich variety of the world as we experience it) is tractable? Newton's answer is, of course, no; according to Newton, time is absolute, not relative.

My point here is not Newton's answer, but the method by which this question is to be addressed. The project of the *Principia* demands the conceptual distinction between absolute and relative time, because (if all bodies are perhaps interacting by means of forces, as yet unknown) there is no material system to which we can turn a priori as a clock. It is then an empirical matter whether there is in fact any material system that acts as a perfect clock relative to the absolute time parameter of the project. Thus, in Newton's hands, the question of whether time is relative or absolute has been turned into an empirically tractable question, and one whose answer depends on the details of empirical enquiry; the appropriate arguments for deciding the question are no longer those based on broadly a priori considerations.

A brief comparison with Aristotle may perhaps be helpful. For Aristotle, time is an aspect of material change, and the motion of the outermost heavenly sphere is the change by which all other changes are measured. In his view, it is not possible to discover empirically that the motion of the heavens (or, equivalently kinematically, the rotation of the Earth) is not uniform and regular. Newton's method leaves the question open. It begins from an assumption of at least approximate uniformity in this rotation, and then harnesses the details in the phenomena to construct a time parameter and to examine the speed of the Earth's rotation with respect to this parameter. As it turns out, it is very hard to tell that the rotation of the Earth (i.e., sidereal time) is not perfectly uniform, that is, that it is not a perfect clock *relative to Newton's time parameter*. It took until the twentieth century to show beyond a doubt that there are tiny irregularities in the rotation of the Earth. But again, it is the mode of argument rather than the conclusion that I want to emphasize. We are not arguing from general principles, or from experience in general, in order to establish our conclusion; rather, we are having to pay attention to the minute details of empirical enquiry in order to address the question of whether there are actual material changes of which time is an aspect, so that these material changes constitute a perfect clock, or whether there are no perfect clocks so that time itself is distinct from material change. It is in this sense that we are doing empiricist metaphysics.

Gassendi's discussion of place, space, and time (in *The Syntagma*; see Gassendi 1972, 383–98) provides another helpful contrast. Newton's treatment bears strong resemblances to Gassendi's discussion, conceptually and in its phrasing (see McGuire 1966). However, two striking differences are pertinent to our purposes. First, Gassendi treats place and space first, and proposes that we can learn about time by drawing on analogies with space. Second, the arguments Gassendi gives are of the general kind that, I am emphasizing, are not to be found in Newton. He argues in general conceptual terms for why time must be prior to and distinct from the motions of material things. It is a methodology of argumentation that Aristotle would have recognized. Gorham (2012) situates Newton's philosophy of time within this tradition of metaphysical arguments for and against the possibility of empty space, in which conclusions about time are drawn by analogy with conclusions about space. Gorham points out that Newton does not argue for absolute time using the kind of metaphysical arguments traditionally offered in support of absolute space. I think that the absence of traditional metaphysical arguments for absolute time in Newton's philosophy of time is because the kind of argument he is offering is different. Newton is offering a methodology in which these questions become empirically tractable, so that a different type of argumentation is required. And I think this represents a big change in philosophy of time. In the wake of Newton's *Principia*, those who accept Descartes's project (as characterized above) have no choice but to develop their philosophy of time, and their account of the relationship between time and material change, in a way that takes into account the types of argument developed by Newton rather than only those that proceed by general conceptual argumentation.

Before moving on, one final comment is in order about the conclusion that Newton draws, that is, that time is absolute. Newton's type of argument does not *prove* that there are no relative motions that constitute a perfect clock, of course. Rather, his argument relies on a conceptual distinction between absolute and relative time, and then provides an empirical argument for the claim that in fact time turns out to be distinct from relative motions of material bodies. And therefore, as such, it is open to empirical refutation, by the finding of a material system that constitutes a perfect clock.[15] But a good empiricist metaphysics proceeds on the basis of empirical evidence and remains open to revision on the basis of future empirical findings, so this is no objection to the claim that, so far as we have been able to ascertain, time is not relative and is distinct from material change.

8.2 *True Time*

Apparent time yields multiple times: apparent solar time, apparent lunar time, and so forth, ticking irregularly with respect to one another. Among the reasons that

---

[15] Moreover, the finding of such a clock would generate further questions about the relationship between that clock and material sub-systems of the universe such that the ticking of the perfect clock could be constitutive of time for these sub-systems.

Gassendi gives against a relationist conception of time is that such an identification of time with actual motions would lead to multiple times. Gassendi asserts that this is to be rejected in favor of a single, unifying time (see Gassendi 1972, 393–94). In Gassendi, the question of whether there is one time or many is to be settled by broadly a priori argumentation, whereas in Newton's hands the issue becomes empirically tractable.

Newton distinguishes between true and apparent time. True time, just like true motion, is a property of the body or system itself, not of the appearances. In the equation of common time, we search for a single time parameter to be used in common for all bodies, relative to which we can construct a theory of their motions. If successful, this time parameter corresponds to the true time of the system of the world. For example, the lunar day and the solar day are different apparent measures of time, ticking irregularly with respect to one another, but belief in true time is the belief that (i) underlying these differing apparent measures is the one true time which these motions approximate to a greater or lesser extent, and *crucially* (ii) the irregularities in the motions of the moon and the sun relative to true time can be *precisely quantitatively* accounted for, without remainder, by our theory of forces by which the bodies in our system interact. The assertion that time is true is the assertion that there is one time, not many, in this specific sense.

True time for Newton, I suggested, is time that is unique and proper to the system of the world. The "system of the world" for Newton is the solar system, but the issue of true time (on the view I advocate) is system-relative and therefore independent of whether the "system of the world" is our solar system or some other system or the material universe as a whole. The issue is whether, for the system of interest, there is any such thing as true time. In the context of Descartes's project, and in the hands of Newton, this is transformed into the following empirical question: can we construct a satisfactory physical theory for this system using a single time parameter?[16]

There are at least two ways in which this could fail. One possibility is that we could get close but not quite there, so that we are left with irregularities that cannot be accounted for either by modifying the time parameter or in terms of forces: we are not able to come up with a satisfactory physical theory using a single time parameter, and so we are unable to supply this empirical warrant for the existence of a true time associated with that system. A second possibility concerns a specific way in which this might happen: subsystems dominated by different forces might tick irregularly with respect to one another with no common underlying metric. For example, there is no guarantee a priori that an atomic clock, governed primarily

[16] Thus, the question also becomes more complex: for any given system, we can ask whether time is true, and also for the collection of all systems we can ask whether time is true.

by the laws of quantum mechanics, will tick regularly with respect to a pendulum, whose rate of ticking is governed primarily by gravitation; the two could turn out to tick irregularly with respect to one another with no common underlying metric. There would then be no such thing as "true time" for any system in which both quantum mechanics and gravitation play a role. Specifically, were it to happen for the solar system, then there would be no such thing as true time for the system of the world.

It is, therefore, an empirical question—and an open empirical question at that—whether any such single time parameter can be constructed. In Newton's hands, the appropriate methodology for tackling the question of whether there is one time or many has been utterly transformed. Whether there is one time or many is something that we discover not through a priori argumentation alone but through attention to the details of empirical enquiry. The question of whether there is one time or many has become an empirically tractable question, one that is appropriately addressed by paying attention to the details of empirical enquiry, and through the progress of that enquiry.

8.3 Mathematical Time

Newton's third distinction between mathematical and common time brings us back to the relationship between time and change. Newton's assertion that time is mathematical includes the claim that it is metrical. What this means is that time is a measure of change: changes occur regularly or irregularly, and faster or slower, with respect to the metric of time, and the comparative length of time between pairs of events is determined by this metric structure. (This is independent of whether time is absolute or relative, and so of whether there are any material clocks, or not. It is also independent of whether time is true or apparent.) As we have seen, time without this much structure seems to be insufficient for the needs of the project of the *Principia*. Thus, the pursuit of the project of the *Principia* is in part an investigation of the claim that the time parameter of the *Principia* must be metrical, and in this way the question of whether time is mathematical has been transformed into an empirical question. Indeed, just how much structure the time parameter must have, in order to fulfill the needs of Descartes's project, is an open question to be explored through the process of empirical enquiry, and it is one that remains an open question today.

There are some interesting conceptual intricacies that arise in the wake of the assertion that time is metrical, and I will finish by offering some comments on these. Following the assertion that time is metrical, there remains the possibility of a gap between time as a measure of change, in and of itself, and time in relationship to actual material changes taking place in the world: why think that the two tick in harmony? Relationists close the gap by identifying units of time with the rhythm of repeating material phenomena, but if time is absolute, as Newton believed, then it is not clear why the ticking of clocks should tell us anything about the metric of time. While time is a measure of change, *we* measure time by means of material change, and if the metric of time were capricious in relation to material change, then material clocks would not measure time. This is not as silly as it sounds: Newton was very alert to the question of whether a unit tick of a clock at one time measures a unit of duration at another time; it is a live possibility for Newton that the length of time between any two ticks of a clock may not be equal. In order to see the significance of this more clearly, it is instructive to consider the difference between Newton's treatment of spatial interval and his treatment of duration.

Spatial intervals are measured by rulers; rulers are bodies, and according to Newton, place is the part of space that a body occupies. A body of unit length by definition occupies a region of space of unit length. There is therefore no distinction between the metrical characteristics of bodies, as occupiers of space, and the metrical characteristics of the parts of space that they occupy. All rulers are, in this sense, perfect rulers: no question arises as to whether a unit ruler measures a unit interval of space, and indeed whether that unit ruler at one location and at one time measures a unit interval of space when moved to another spatial location and/or at another time.

By contrast, not all clocks (indeed perhaps no clocks) are in an analogous sense perfect clocks. And this is the point about the possible capriciousness of time: a unit tick of a clock at one time may not measure a unit of duration at another time. For Newton, there is a gap between duration and our measure of duration that does not arise for length and our measure of length. This is why Newton need not specify explicitly that space is mathematical, whereas he must—and does—do so for time.[17]

The gap between duration and our measure of duration therefore presents us with an epistemic problem, in that we cannot measure duration directly, but only indirectly by means of motion. Closing this gap involves two steps. First, we have to specify that time is metrical, and this is something Newton does. Then, we have to establish a relationship between this metric of time and the "ticking" of material clocks, so that our means of measuring duration is not utterly unreliable as a guide to the metric of time. Newton does this by stipulating that time "flows equably": the metric of time is not capricious in relation to physical processes.[18] To say that time

[17] Moreover, this marks a difference as compared to mass, as a measure of matter, and momentum, as a measure of motion, for example, where Newton is able to formulate quantities that do not suffer from the same kind of epistemological gap. In the case of time, it seems that no amount of conceptual work will remove the gap.

[18] Arthur (1995) argues, though for different reasons, that we should regard Newton's assertion that time is mathematical as asserting that it flows equably, as well as that it is metrical.

"flows equably" is to say not only that it is a measure of change, but that the processes and changes that we experience as regular *are* regular, at least approximately, with respect to time. Were it otherwise, then the metric of time would be epistemically utterly inaccessible to us.

Thus far, Newton has been making clear what demands are placed on our concept of time by the requirement that material clocks enable us to measure intervals of absolute time. The *evidence* that time flows equably is the practical success achieved by astronomers with the equation of time, and by Huygens with the pendulum clock. There may be no perfectly equable motions to be found in the material world, since there may be no material clock that ticks precisely in accordance with the metric of time, but we have good empirical reasons in support of the claim that absolute and true time flows equably.

I think that the issues here are more complex than those raised by absolute and true time, and that the subsequent history of work on space and time bears this out. The relationship between material processes and space and time continues to be rich philosophical territory in foundations of physics, in which a wide array of positions is available. There are disputes over the priority of space and time versus matter, over the minimal structure that we must attribute to space and time, over the nature of the relationship between space-time and matter, and of how it is that rods and clocks, such complex material systems, overcome that complexity to somehow tell us something about the structure of the spacetime in which they are situated and move. It is not at all obvious how to resolve these questions empirically, and what this makes vivid is the sophisticated reasoning that is involved in bringing the empirical to bear on such difficult philosophical questions. Nevertheless, what is also clear is that it is through the process of this very engagement with the details of empirical enquiry that progress on these questions is to be made. Again, the methodological point is the one that I wish to emphasize, not the answer to the question of whether or not time is mathematical. It is because of the moves that Newton makes that we uncover all the philosophical complexity associated with this claim in the first place.

9. CONCLUSIONS

I have argued that Newton is very deliberate when he sets out his three distinctions between absolute and relative, true and apparent, and mathematical and common time: each distinction is different, and each is needed for setting up the project of the *Principia*. I have also argued that all three distinctions mark open empirical questions in the context of the *Principia* (though I make no claim that this was Newton's own view). The questions of whether time is absolute or relative, true or apparent, and mathematical or common become empirically tractable in the context of the project of the *Principia* (or some such project). Moreover, the significance of this conclusion is not local to the *Principia*. If we read the *Principia* as making a contribution to Descartes's "metaphysical physics," then we can read the moves that Newton is making as contributions to that metaphysical project, albeit via a very different epistemology and methodology. Read in this light, we see that Newton's distinctions are connected to questions concerning the nature and structure of time. Newton transformed the methodology by which these questions should be addressed, providing empirical purchase on them and rendering them empirically tractable. Prior to Newton, questions such as whether time is relative (depending on the actual motions of bodies) or absolute (independent of these motions), and whether there is one time or many, were questions that could be tackled independently of the details of empirical enquiry. In the wake of the *Principia*, this is no longer the case: any legitimate exploration of these questions must take into account the kind of detailed empirical enquiry pursued in the *Principia*. Philosophical progress has been made.

In addition to these specific conclusions concerning an appropriately empiricist metaphysics of time, there is a more general methodological message that is perhaps worth highlighting. Disputes in philosophy of time over how to understand the significance of special relativity, for example, are hampered by difficulties over how to read the developments in physics as contributions to philosophy. Disputes arise as to whether the philosophical challenges posed by relativity of simultaneity in special relativity can be safely ignored by philosophers since special relativity is a false theory, and over whether, even if special relativity is to be taken into account, the dispensability of a privileged present in special relativity implies anything about the existence or otherwise of a privileged present, metaphysically.[19] These disputes seem to me poorly framed, because they fail to read special relativity itself as arising through a diachronic process of philosophical engagement with our everyday concepts of space and time, a process that has clarified and transformed those concepts (see DiSalle 2006). The example I have discussed in this paper is Newton's work on duration, and I have said nothing about simultaneity, which is more of a hot topic in contemporary metaphysics. But similar work to that which I have done here can be done for simultaneity. The question of whether time flows is another topic that has received attention recently, and, as my remarks on mathematical time above indicate, I think that attention to the role of this claim in Newton's project help clarify how such a claim should be understood. By reading developments in physics as a part of the history of philosophy, we can make visible the ways in which physics

[19] For a recent example where these issues arise, see Zimmerman 2013.

contributes to the conceptual clarifications and transformations of the very questions that we are asking about time, and the means by which aspects of our oldest questions concerning time are rendered empirically tractable.[20]

The *Oxford Handbook of Philosophy of Tim*e (2011) and the *Blackwell Companion to the Philosophy of Time* (2013) both contain sections written by metaphysicians (on such topics as presentism, persistence, fatalism, and tense) and sections written by philosophers of physics (on time in classical and relativistic physics, and in cosmology and quantum gravity). Though I have little empirical evidence to offer you, I am doubtful how much each group of philosophers reads one another's chapters: if cross-referencing is anything to go by, they are not really talking to one another, and one reason is their differing methodologies. A more historically driven methodology, in which both philosophy and physics are read diachronically as contributions to our ongoing philosophical conversation, would enable both parties to speak to one another in more fruitful ways.

ACKNOWLEDGMENTS

I am grateful to the participants in my graduate spacetime seminar (spring 2013), and to our visiting speaker Eric Schliesser. My thanks to the Pittsburgh Center for Philosophy of Science, the Stanford Philosophy Department, and the London School of Economics, and to all those present at my talks. Special thanks to Ori Belkind, Meghan Dupree, Jennan Ismael, Anja Jauernig, Xavi Lanao, Tom Ryckman, George Smith, Monica Solomon, Michael Strevens, and Zanja Yudell for their comments and discussion.

REFERENCES

Arthur, R. T. W. 1995. "Newton's Fluxions and Equably Flowing Time." *Studies in History and Philosophy of Science* 26 (2): 323–51.

Audoin, Claude, and Bernard Guinot. 2001. *The Measurement of Time*. Cambridge: Cambridge University Press.

Callendar, Craig. 2011. *The Oxford Handbook of Philosophy of Time*. Oxford and New York: Oxford University Press.

Descartes, R. 1991. *Principles of Philosophy*. Translated by V. R. Miller and R. P Miller. Dordrecht, Boston, London: Kluwer.

DiSalle, Robert. 2006. *Understanding Space-Time*. Cambridge: Cambridge University Press.

20 See, for example, Smeenk (2013), who discusses the transformation of old questions by developments in cosmology.

Dyke, Heather, and Adrian Bardon. 2013. *A Companion to the Philosophy of Time*. Blackwell Companions to Philosophy. Chichester, West Sussex: John Wiley & Sons.

Garber, Daniel. 1992. *Descartes' Metaphysical Physics*. Chicago and London: The University of Chicago Press.

Gassendi, P. 1972. *The Selected Works of Pierre Gassendi*. Edited and translated by C. B. Brush. New York and London: Johnson Reprint Corporation.

Gorham, G. 2012. "'The Twin-Brother of Space': Spatial Analogy in the Emergence of Absolute Time." *Intellectual History Review* 22 (1): 23–39.

Hall, A. R., and M. B. Hall, ed. 1962. *Unpublished Scientific Papers of Isaac Newton*. Cambridge: Cambridge University Press.

Huggett, N. 2012. "What Did Newton Mean by 'Absolute Motion'?" In *Interpreting Newton: Critical Essays*, edited by A. Janiak and E. Schliesser, 196–218. Cambridge and New York: Cambridge University Press.

Janiak, Andrew. 2008. *Newton as Philosopher*. Cambridge: Cambridge University Press.

McGuire 1978. "Existence, Actuality and Necessity: Newton on Space and Time." *Annals of Science* 35: 463–508.

McGuire, J. E. 1966. "Body and Void in Newton's *De Mundi Systemate*: Some New Sources." *Archive for the History of Exact Sciences* 3: 206–48.

Newton, I. 1999. *The Principia: Mathematical Principles of Natural Philosophy*. Trans. I. B. Cohen and A. Whitman. Los Angeles: University of California Press.

Palmerino, C. R. 2013. "The Composition of Space, Time and Matter According to Isaac Newton and John Keill." In *The Mechanization of Natural Philosophy* (Boston Studies in the Philosophy of Science), edited by D. Garber and S. Roux, 117–42. Springer.

Schliesser, E. 2013. "Newton's Philosophy of Time." In *A Companion to the Philosophy of Time*, edited by H. Dyke and A. Bardon, 87–101. London: John Wiley and Sons, Inc.

Smeenk, C. 2013. "Time in Cosmology." In Dyke and Bardon, 2013.

Stein, H. 1970. "Newtonian Space-Time." In *1666: The Annus Mirabilis of Sir Isaac Newton*, edited by R. Palter, 258–84. Cambridge, MA and London: MIT Press.

Zimmerman, D. 2013. "Presentism and the Space-Time Manifold." In *The Oxford Handbook of Philosophy of Time*, edited by Craig Callendar, 16–244. Oxford and New York: Oxford University Press.

2

Ontology, Complexity, and Compositionality

Michael Strevens

A CERTAIN APPROACH to the metaphysics of science, or to scientific metaphysics, is demonstrated in this piece about compositional theories of complex systems. In these opening remarks I describe the approach; my aim in the rest of the paper is not to argue for so much as to illustrate what I have described.

My specialty is the metaphysics of the higher level—the way that the fundamental constituents of the material world and the fundamental laws come together to compose or otherwise determine the existence of high-level causal relations, probability distributions, laws, entities, properties, and other manifestations of non-fundamental being.

I take the basic physical stuff as given, then—whether it is provided by a Humean (Lewis 1994), a realist (Armstrong 1997), a scientific essentialist (Ellis 2001), or some other metaphysical *Bauplan*. My question is what else exists above and beyond the fundamental level, in virtue of the fundamental level, and why. Do high-level laws exist in virtue of their entailment or probabilification by the fundamental-level laws (Loewer 2008)? Or in virtue of their being systematizers par excellence of singular high-level facts (Cohen and Callender 2009)? Or in virtue of a special mix of physical necessity and singular fact that gives them the ability to support a wide range of counterfactuals (Strevens 2008b)? These are my kinds of problems.

The key to solving such problems is, I hold, scientific explanation: the high-level laws, probabilities, and entities are endowed with ontological significance by their role in making sense of the world. Science carves them out of the fundamental-level substrate to take advantage of their explanatory power, rather than to admire their metaphysical curves. That explanatory carving itself, however, confers metaphysicality: the answer to the question, "What are the high-level laws?" is "Whatever plays the law role in high-level explanation." Likewise, the answer to the question, "What are the high-level entities?" is "Whatever plays the entity role (or roles) in high-level explanation."

I have given a sketch of the resulting metaphysics of special science laws, deterministic and statistical, elsewhere (Strevens 2008a, §§7.6 and 10.5). Here, I hope to do the same for a certain specific class of physical probabilities. Although these probabilities are not natural from a naive or *a priori* perspective, they play an important explanatory role in science. In virtue of their power to illuminate the world, I propose, they deserve a place in the world.

I offer no argument for this proposal. Rather, I hope that the picture of the high-level world implied by the view is sufficiently attractive to command respect and even allegiance on its own merits.

My method is straightforward. The inventories of entities compiled by the various sciences for explanatory purposes, I simply take for granted as genuine ontologies. What the sciences say exists—provided that it has a genuine theoretical function—does exist. You can see for yourself whether or not you like the consequences.

1. THE WEDDING CAKE ONTOLOGY

Different sciences have different ontologies—different ways of dissecting the world into individuals, categories, properties. Fundamental physics does particles, chemistry does molecules, biology does cells and organisms and ecosystems, and so on. The list suggests that a certain neat structure is the rule in this grand ontological project: the things at one level are spatiotemporally composed of the things at the next level down. Animals are made of cells, which are made of molecules, which are made of particles . . .

It follows that the various sciences will cooperate in a straightforward way. Physics, in determining how particles behave, will determine also how spatiotemporal complexes of particles behave—and these rules are the laws of chemistry. Chemistry, in determining how molecules behave, will determine also how complexes of molecules behave—and these rules are some of the laws of geology, cytology, and so on. Each science builds its subject matter from the parts supplied by the science below, and deduces its laws from the laws of those lower-level parts, also supplied by the science below. This wedding cake picture of the unity of science was made famous by Oppenheim and Putnam (1958); I am here applying the metaphor to its ontological component.

Why should the wedding cake be the rule? One advantage of the cake's structure is that it promises to supply to the scientific enterprise a copious array of *compositional theories*. A compositional theory earns that sobriquet by providing a tool kit for understanding a wide range of systems within a given class (using the term *system* broadly to encompass molecules, organisms, ecosystems, and minds). It divides the system into parts and assigns dynamical properties to these components that, when aggregated, predict the behavior of the system as a whole. If you have a compositional theory of a certain kind of system, then, you do not need to theorize anew for each instance of that system. You have a procedure you can use to predict and explain the behavior of every configuration of the system within the allowed range.

Warning: compositionality of theories is not the same thing as composition of objects. The one is a property of a system of representations, in virtue of which the behavior of the whole is derived from, and hence predicted and explained by, the behavior of the parts; the other is a metaphysical relation, in virtue of which the existence of the whole is secured by the existence, properly arranged, of the parts. This paper talks about both compositionality and composition; there is, however, no straightforward relation between the two.

Compositional theories are, quite obviously, greatly desirable; indeed, it is hard for science to make much headway against the world's complexity without them. A compelling attraction of the wedding cake picture is that it yields a template for building compositional theories. It is spatiotemporal decomposition: divide your system into spatiotemporal parts; then apply the rules that govern the parts, aggregating to deduce the behavior of the whole.

As I will argue in the next section, however, spatiotemporal decomposition frequently leads to a complexity catastrophe: due to the sensitivity and combinatorics of the interactions between parts, it is impossible in practice to aggregate the behavior of the parts to obtain a description of the behavior of the system that they compose.

The solution is to put aside the wedding cake's spatiotemporal ontology and to find other ways to build compositional theories. In this paper I examine one such approach, a statistical route that might seem at first to help itself to a traditional wedding cake taxonomy, but that on closer examination slices up the systems to which it applies in a way that is not at all spatial, and that consequently cuts across the wedding cake division of the world into parts.

The result is quite compatible with reductionism: it is physicalist and fundamentalist, meaning that everything is made up of physical stuff and everything happens because of the ways that the fundamental laws of nature push around such stuff. The bottommost layer of the cake is, then, firmly in place. From there on up, however, the theoretical patisserie takes on a variety of stranger and more intricate forms. The familiar layers are inarguably there—molecules, cells, organisms, and the rest—but there is much else squeezed into the interstices.

2. COMPOSITIONALITY AND COMPLEXITY

Choose a class of systems: halide molecules, coronaviruses, predator/prey ecosystems. A compositional theory of such a class divides the systems into parts, assigns the parts properties in virtue of which they behave in certain ways, and then aggregates the behavior of the parts to derive, for predictive or explanatory purposes, the behavior of the whole—for any system in the class.

Newtonian gravitational theory, for example, divides the world into objects, assigns those objects masses in virtue of which they exert and experience gravitational force, and then by way of a principle of aggregation—the rule that individual forces are to be summed as vectors—makes predictions about the movements of a system of such objects.

In principle, Newtonian theory can predict the behavior of any system in its scope; there lies its compositional strength. In practice, great difficulties may emerge in the aggregation. Given only three gravitating bodies, the mathematics of even approximate derivations can, if the bodies are of similar mass, become complex, in part because such systems may be highly sensitive to initial conditions. For greater numbers of bodies, combinatorics piles on difficulty: a small change here can make for moderate changes in many other places, which in turn change things further here—and so on. This is a complexity explosion; it limits the usefulness of even so practically important a compositional theory as Newtonian gravitation.

What you might call the *aggregation problem* arises again and again at many different levels in the sciences. The quantum chemistry of large atoms is difficult enough; that of large molecules is more challenging still. Modeling the complex genetic networks at work in embryological development is fiendishly hard. Predicting many of the significant consequences of interacting human minds—housing bubble collapses, Hollywood megahits, popular revolutions—is quite beyond us. Sometimes, however, compositional theories of complex systems lie within our grasp: we can build successful models in statistical physics, evolutionary biology, and computational psychology. What many of these models have in common is their being non-spatiotemporal in some aspect of their underlying ontology: they decompose systems into parts or properties, some of which are not spatiotemporally discrete units or anything close. The need to solve the problem of aggregation, then, imposes on higher-level sciences ontologies that depart at least in part from the wedding cake ideal.

Is this because theories with a wedding cake ontology are especially prone to the aggregation catastrophe? Is there something about decomposing a system into spatiotemporal parts that renders the calculus of aggregation especially intractable? Perhaps not. It might simply be that the great majority of lower-level ontologies are unsuited to tractable aggregation, and that since spatiotemporal cohesion is a strong constraint and so wedding cake ontologies are only a small proportion of the possible ontologies, we ought not to have expected them to be particularly successful.

Since the spatiotemporal constraint is chosen not at random but because it simplifies the organization of science, that is a somewhat weak explanation of the aggregation problem. I think we can do better: the nature of the wedding cake ontology to some extent does explain why it, in particular, suffers from the aggregation problem.

A rule of aggregation pulls together the connections between the parts into which a system is decomposed. In a compositional theory with a wedding cake ontology, then, the aggregation rule will calculate the net effect of the relevant connections—causal connections, let me suppose—between its spatiotemporal parts. The aggregation problem tends to arise for wedding cake theories because of the combinatorial complexity and sensitivity of these relations.

I call the relations between spatiotemporal parts *sensitive* because their effect on the aggregate is sensitive to small changes in the state of the part. I do not mean "sensitive" in the chaotic sense, but in a much weaker sense: the effect on the aggregate is not wholly independent of the small changes, or in other words, small changes have some effect on the aggregate. For example, although the gravitational force exerted by an object does not depend chaotically on the object's position, it does depend on exact position: move the object slightly and the force exerted at any point changes slightly, and so the aggregate force exerted on an object at that point changes slightly. The same is true for mass: a slight change in mass means a slight change in force, rather than none whatsoever.

The relations between spatiotemporal parts are *combinatorially complex* because, as you increase the number of objects, the number of relations to keep track of increases. I do not mean that it increases exponentially, or even non-linearly—just that there is an increase.

These two properties tend to lead, because of increasingly intricate whorls of dependence, to complexity explosions: as in the case of gravitation, small alterations in one place proliferate quickly to many other places, where they further nudge conditions at the original locus of change. The web of mutual influence becomes a hopeless tangle.

For this reason, wedding cake ontologies typically prove to be unsuitable ground for compositional theories of complex systems. What theories do better?

3. ENION PROBABILITY ANALYSIS

A broad class of compositional theories dissolve the aggregation problem by taking a statistical approach: they divide the system into individuals such as molecules or organisms (which I call "enions"), they assign probability distributions to the behavior of the enions, and they derive the behavior of the whole by aggregating the relevant probabilities. I call this strategy *enion probability analysis*, or EPA for short (Strevens 2003).

Enion probability analysis cannot be used to model just any property of a complex system: it is limited to tracking aggregate properties of populations. You can use it (when conditions are favorable) to model changes in the number of tortoises in an ecosystem, for example, or the number of male tortoises in the system between one and two years of age, but not to follow the day-to-day movements of a particular tortoise. (For the latter purpose, I know of no compositional theory that does not suffer from the aggregation problem.) This is a relatively small subset of the properties of a system, but it is an important one, containing the things we need to predict and explain in order to undertake many of the projects falling within the ambit of statistical physics, evolutionary biology, ecology, economics, and so on. Consequently, EPA has a significant role to play in accounting for the predictive and explanatory success of the compositional sciences of complex systems, and a philosophical understanding of EPA has significant lessons to teach about the ontologies of successful compositional theories—and thus, given my explanationist approach to these issues, about the ontology of the world.

In the boreal forest of Canada, snowshoe hares eat underbrush and Canadian lynxes eat snowshoe hares. The populations of the two species cycle regularly: every eight to eleven years, the hare population booms (for reasons about which there is still considerable argument), and the lynx population, sustained by regular helpings of *lapin à la mode*, booms soon after. Then the hare population crashes and the lynx population follows on its heels.

Ecologists wishing to understand this famous predator/prey population cycle build mathematical models of the system. Although the models themselves tend to have a deterministic mathematics, their structure can be understood as rooted in stochastic foundations, as it was by Alfred Lotka, who gave his name to these "Lotka-Volterra" models (Lotka 1925). This is the statistical method that I am calling enion probability analysis; after describing its application to lynxes and hares, I will examine its presumptive ontology.

A Lotka-Volterra model tracks populations, attempting to predict the size of future populations—say, the numbers of lynxes and hares in a month's time—from present populations. Some sophistications are possible: weather patterns or the luxuriance of the vegetation can be incorporated into the model, or subpopulations can be tracked (such as sexually mature female hares). But let me put all of this aside to focus on the basics.

Assume that the model has only two variables, representing the number of lynxes and the number of hares. What is wanted is a set of equations that relates the values of the variables at one time to their values at a later time.

It might seem that a complexity explosion is imminent. What happens to an individual hare depends on small wrinkles in local matters of fact. Forage under the old pine this morning and you will be lynx food. Head instead for the spruces, and you will live to graze another day. Tracking the fates of individual hares requires, then, an accurate record of particular movements and the principles that drive them. That, surely, must add up—when hundreds or thousands of hares and many lynxes are involved—into something very complex, certainly more complex than a set of simple equations relating only two variables. Or so it would appear.

Lotka's approach gets around these difficulties using the following recipe. To the members of each population or sub-population in the ecosystem, assign probabilities for the kinds of outcomes that make a difference to populations: birth, death, perhaps migration. Assume that these "enion probabilities" are independent of one another. Use the law of large numbers—assuming here that populations are reasonably large—to derive frequencies for various events equal to the enion probabilities. If, for example, there is a 0.05 probability that any given hare is killed by a lynx over the course of a month, assume that one-twentieth of the hare population is lost to lynx predation every month. You now have a table of per capita rates: the rate of hare reproduction (per hare), the rate of hare predation, the rate of lynx reproduction, the rate of lynx death, and so on.

Some of these rates depend on other variables. The rate of hare predation depends, for example, on the population of lynxes: more lynxes means proportionally more hares served up for dinner. These dependences typically (though not invariably) end up in the model. It is crucial, then, that they bring with them into the model's equations only quantities that the model is constructed to represent. In the simple case at hand, the rates should depend only on the total number of hares or the total number of lynxes in the system, or both, or neither. (If the rates as a matter of fact depended on subpopulations, the model would have to track those subpopulations—young female hares, for example—thereby becoming more complex.) The rates will satisfy this independence requirement just in case the enion probabilities satisfy the requirement. The probability of a hare's being killed by a lynx over the course of a month should, for example, depend only on the number of lynxes (and perhaps the number of hares).

Let me suppose that the necessary enion probabilities, and so the rates, are fully determined by the physical facts. (Strevens [2003] tries to specify exactly what fundamental states of affairs fix the facts about the probabilities.) Then, with the probabilities' existence secured, it is straightforward to write down two equations representing, respectively, the rate of change of the hare and lynx populations. The equation for the hare population might, for example, set the change in the population equal to the rate of hare reproduction (multiplied by the current hare population) less the rate of predation (multiplied by the current hare and lynx populations) less the rate of hare death from other causes (multiplied by the current hare population). The lynx equation will do something similar. Together, these equations make up a Lotka-Volterra model, which can be simulated or solved and then either tested against observed population changes or used to explain those changes.

That is one way—the EPA way—to model a complex system, deftly avoiding the problems posed by the complexity explosion.

The enion probability analysis of the lynx/hare system appears at first to have a wedding cake ontology: it proceeds by dividing the system into enions, individual hares and lynxes, that are spatiotemporally discrete. A closer look shows, however, that although the enions are indispensable, the most important elements of EPA—not the enions themselves, but the probabilities that describe the enions' behavior—are individuated rather differently.

4. WHERE DID THE INTERACTIONS GO?

An explosion was averted. Every hare and every lynx has manifold interactions with its environment; changes in the hare and lynx populations are nothing but the aggregate consequences of these interactions; and aggregating the interactions looks to be the sort of intractable task that threatens to sink the sciences of complex systems—yet EPA proceeds without a hitch. What happened to the interactions? How were they, in the end, so easily aggregated? By answering this question in part, I will show you that the ontology of EPA does not conform to the model of the spatiotemporally organized layer cake.

Explaining the complexity explosion in wedding cake models above, I attributed it to two properties of the relations between the parts of those models: sensitivity and combinatorial complexity.

An inter-part relation is sensitive if small changes in the state of a part make for a difference (perhaps slight) in the relation. It gives rise to combinatorial complexity if the number of relations the model must keep track of (when aggregating) increases with the number of parts.

In a Newtonian model, the relevant relations are the connections that determine the force exerted by a spatiotemporal part—the forces that must be aggregated, that is, to determine the behavior of the system as a whole. In an EPA model, they are the relations that determine enion probabilities such as the chance of hare death—the probabilities that must be aggregated to determine the behavior of the system as a whole.

Unlike the Newtonian relations, the EPA relations are not sensitive: enion probabilities are not affected by small changes in the state of the relevant enion, such as a shift in position. Indeed, by the independence requirements stated above, they depend on almost nothing about any enion—the probability of a hare's death is not affected by its position or by what happens to any other hare or lynx.

Nor are the EPA relations combinatorially complex: the enion probability analyst must keep track of one set of probabilities per enion type, and that is all. In the lynx/hare system that amounts to two sets, one for lynxes and one for hares, regardless of the population of each. A system with many lynxes is consequently no more difficult to model than a system with a few. Indeed, large populations make things simpler, by making it more likely that actual behavior will correspond to statistically expected behavior.

These negatives—the lack of sensitivity and of combinatorial complexity—go some way toward explaining why EPA models do not suffer from a complexity explosion, but there is much more to be said. The source of enion probabilities' insensitivity is particularly important: the key to understanding the power of EPA is, I think, to understand why there is so little dependence between the statistical behavior of enions and their exact or even approximate states. To put it another way, what should be explained are the independence assumptions upon which the applicability of EPA depends:

1. Enion probabilities depend only on population-level variables of the sort tracked by statistical models.
2. The outcomes to which enion probabilities are attached are stochastically independent.

This is a project I tackle in Strevens (2003); an overview is given in Strevens (2005). The complete story is not something that I will undertake to give here. For the purposes of understanding the implications of compositionality for ontology, it will be enough to answer an easier question: where, in EPA, do the interactions go?

We know that there are many interactions between hares and other hares, lynxes, and their environment. Try to track these many interactions and you will generate—so I have supposed—an immediate combinatorial catastrophe. Enion probability analysis, by representing the behavior of the system as a whole, represents the aggregate effect of these interactions. Yet it somehow, in its formalism, avoids having to represent the interactions explicitly—and so avoids having to aggregate them formally, bypassing the aggregation problem that would result. The interactions are packed away in some place in which they cannot get out of hand. Where?

Consider the probability of a hare's being killed by a lynx over the course of a month—the number that determines (more or less) the rate of hare predation. On what features of the lynx/hare ecosystem does the value of this number depend? What aspects of the system go into determining that the probability of hare death per month is, say, 0.05 rather than 0.1? The relevant factors include the total number of lynxes, the techniques that lynxes use to hunt hares, the techniques that hares use to avoid lynxes, the nature of the vegetation in the habitat, and more. Change any of these things in significant ways, and the magnitude of the probability of hare death will surely change.

The dependence of the hare death probability on the first of the enumerated factors—on the number of lynxes—is represented explicitly in the EPA model. The effect of lynx number on the probability is, in other words, "externalized." What about the rest? They are entirely internal to the probability, which is to say that their net effect is built into the probability—in formal terms, built into the 0.05; in metaphysical terms, built into the physical probability quantified by that number. As a consequence, the model need not explicitly take these interactions into account.

There I will pretty much leave the explanation of the miracle of EPA, taking away two claims about enion probabilities. First, the probabilities are not physically separate and independent entities. They are attached to physically independent entities—to different hares—but they physically overlap, since numerically identical states of affairs contribute to many distinct probabilities. The lynxes' tactics, for example—ultimately a matter of lynx brain configuration, I suppose—help to determine each hare's probability of death, as do many other shared aspects of lynx makeup. Follow the death probabilities for different hares down to the fundamental level, then, and they converge on many of the same fundamental-level facts. In other words, the reduction or supervenience bases for any two hares' death probabilities overlap. Contrast this with a standard wedding-cake theory such as Newtonian gravitation, on which each object contributes to and experiences the net gravitational field in virtue of a wholly spatiotemporally intrinsic property, its mass. The principles for enion probabilities' individuation bear little resemblance to the principles for the construction of the wedding cake.

Second, it is this extrinsic and overlapping quality that makes it possible for EPA to avoid an explosion of complexity, opening the door to a compositional theory that shrugs off the aggregation problem. How so? I venture that a compositional theory must, in order to be useful, individuate a system's parts and properties so that they are in some sense largely independent. Enion probability analysis does not, in its delineation of the determinants of enion behavior, divide the world into factors that are physically independent. It does, however, divide the world into determinants of behavior that are *stochastically* independent, and here lies its power: the rules for aggregating stochastically independent determinants of behavior are far more tractable than the rules, in any interesting system, for aggregating physically independent determinants.

A metaphysical postscript: I have assumed that we find in the fundamental-level world all the materials, factual and nomological, that we need to build enion probability distributions, and I have shown that science uses the distributions so constituted to predict and explain a great many things. Let me now assert, as promised in the prefatory note to this paper, that this predictive and explanatory prowess confers on the probabilities a commensurate ontological status: they are real probabilities (Strevens 2011), and their individuating principle therefore traces a real ontological fault line.

5. ONTOLOGIES OF THE HIGH-LEVEL SCIENCES

The high-level sciences are ontologically pluripotent. They make great use of physical individuation, that is, individuation according to the spatiotemporal wedding cake conception: even in EPA, the bearers of probability are typically physically discrete entities such as molecules, animals, or people, and the ultimate aim of models is to track the statistical movements of such things.

Yet at least as important are what you might call distributed ontologies, that is, individuations into entities—things and reifications of the tendencies of things, such as causal dispositions and enion probabilities—whose presence is determined by configurations of fundamental physical facts that overlap, so that the same facts contribute to many such entities. Enion probabilities are this paper's paradigms of distributed ontology.

The entities distinguished by the first, spatiotemporal kind of ontological decomposition—the enions themselves, the lynxes and hares—are physically independent but interact with one another in many complex ways. The entities distinguished by the second, distributed kind of ontological decomposition—the enion probabilities, such as the probability of a certain hare's death over the course of a month—are physically overlapping, but stochastically independent and therefore easily aggregated.

These two ontological schemes are not rivals, but rather work together within a single modeling technique in population ecology, serving up a compositional theory that solves the problem of aggregation. The wedding cake ontologists are right to think that spatiotemporal individuation has been essential to creating compositional theories of the high-level sciences, but wrong in thinking that it has been sufficient. To tackle the sciences of complex systems, we need what is, in a mild sense, ontological pluralism.

Can the same be said of other kinds of compositional theories? Let me give you two examples.

The first is spectral analysis in wave theories of various aspects of nature. In the high-level sciences, there are sound waves, ocean waves, waves on the strings of musical instruments, seismic waves, and more—where in each case, the wave is a movement of an underlying medium. In spectral analysis, the medium's movement is decomposed into waves of different frequencies, as when the motion of a vibrating string is decomposed into a tone and various overtones. These waves coexist in the same medium—in the same volume of air, or earth, or water—and indeed in the same movements of that medium; consequently, the fundamental-level matters of fact on which different waves in a spectral decomposition depend are largely identical. The waves form a distributed ontology.

For the predictive and explanatory purposes of many wave models, the only aggregation required is the addition of the effects of these different frequencies, which is accomplished by the straightforward process of linear superposition. The wedding-cake alternative, in which a model keeps track of the movements of different parts of the medium—different segments of a vibrating string, or different volumes of air or water—is far more difficult to implement. A distributed ontology brings compositional modeling within reach.

(Our theories of the forces that come together to create waves are, however, often sensitive and combinatorially complex. As a result, wave theories will, for certain predictive purposes, suffer from a complexity explosion. That is why quantum chemistry is computationally so difficult.)

Another, more speculative, example is belief/desire psychology. Most of us would guess that the facts underlying beliefs and desires—the facts that make it the case that I believe that there is rabbit for dinner or that I desire to wash it down with a glass of wine—are to some extent distributed across the brain in an overlapping way. The propositional attitudes comprise a distributed ontology.

Belief/desire psychology is also a remarkably effective compositional theory of thought (Dennett 1987). The principles of composition are quite familiar to us, but the relation of the whole to the underlying facts remains, for now, opaque. It seems that belief/desire psychology solves an aggregation problem by way of a distributed ontology, then, but we cannot as yet be sure.

Some philosophers have suggested that beliefs and desires not be taken ontologically very seriously at all (Churchland 1981; Dennett 1987), in part, I would guess, because their distributed nature lends them a certain insubstantiality, a lack of proper placement within the great wedding cake of science. Perhaps contemplation of the role of distributed ontologies elsewhere, in wave theories and in EPA, can solidify the attitudes' status, both scientific and metaphysical.

REFERENCES

Armstrong, David M. 1997. *A World of States of Affairs*. Cambridge: Cambridge University Press.

Churchland, Paul M. 1981. "Eliminative Materialism and the Propositional Attitudes." *Journal of Philosophy* 78: 67–90.

Cohen, Jonathan, and Craig Callender. 2009. "A Better Best System Account of Lawhood." *Philosophical Studies* 145: 1–34.

Dennett, Daniel. 1987. *The Intentional Stance*. Cambridge, MA: MIT Press.

Ellis, Brian. 2001. *Scientific Essentialism*. Cambridge: Cambridge University Press.

Lewis, David. 1994. "Humean Supervenience Debugged." *Mind* 103: 473–90.

Loewer, Barry. 2008. "Why There *Is* Anything Except Physics." In *Being Reduced: New Essays on Reduction, Explanation, and Causation*, edited by Jacob Hohwy and Jesper Kallestrup, 149–63. Oxford: Oxford University Press.

Lotka, Alfred. J. 1925. *Elements of Physical Biology*. Baltimore, MD: Williams and Wilkins.

Oppenheim, Paul, and Hilary Putnam. 1958. "Unity of Science as a Working Hypothesis." In *Minnesota Studies in the Philosophy of Science*, vol. 2, *Concepts, Theories, and the Mind-Body Problem*, edited by Herbert Feigl, Michael Scriven, and Grover Maxwell, 3–36. Minneapolis: University of Minnesota Press.

Strevens, Michael. 2003. *Bigger Than Chaos: Understanding Complexity through Probability*. Cambridge, MA: Harvard University Press

Strevens, Michael. 2005. "How Are the Sciences of Complex Systems Possible?" *Philosophy of Science* 72: 531–56.

Strevens, Michael. 2008a. *Depth: An Account of Scientific Explanation*. Cambridge, MA: Harvard University Press.

Strevens, Michael. 2008b. "Physically Contingent Laws and Counterfactual Support." *Philosopher's Imprint* 8 (8): 1–20. http://quod.lib.umich.edu/p/phimp/3521354.0008.008.

Strevens, Michael. 2011. "Probability out of Determinism." In *Probabilities in Physics*, edited by Claus Beisbart and Stephan Hartmann, 339–364. Oxford: Oxford University Press.

3

Naturalized Metaphysics and the Contention over the Ontological Status of Species

Matthew H. Slater

I. NATURALIZED METAPHYSICS AND THE METAPHYSICS OF SPECIES

In their classic text on plant evolution, botanists Briggs and Walters asked, "When we look at nature, are the 'units' we recognise and name already there to be recognised or have we 'made' them in the process of looking?" (1997, 361). Are species, in other words, *real*?[1] Assuming that species *are* real and there to be discovered in advance of our efforts to name and characterize them, what sorts of things are they? As Michael Ruse once asked, are they "natural kinds, individuals, or what?" (1987).

Such questions concerning the metaphysics of species raise issues that are paradigm instances of scientific metaphysics (or as Kyle Stanford might prefer, "the metaphysics of science"; see chap. 7 of this volume). As such, the general topic seems a fine test case for proposals on how best to "naturalize" metaphysics, particularly if considering them might help us get more precise about what naturalized metaphysics is and why it is supposed to be more valuable than non-naturalistic (or "analytic" or "speculative") metaphysics. Yet no one (to my knowledge) has explored the relevance of the species-metaphysics debate to broader questions about the status

[1] As Coyne and Orr point out, "Most biologists certainly *act* as if species are real. . . . Yet a vocal group of biologists, including many botanists, dissent, claiming that species are subjective divisions of nature made for human convenience" (2004, 9).

of naturalized metaphysics debate in any detail. This is somewhat surprising given that arguments in these debates appeal to explicitly naturalistic considerations. Defenders of the dominant metaphysics of species, the Species-as-Individuals (SAI) thesis, often center their arguments for SAI on its tight relationship with common practices in systematics—in purported contrast with competing views that diverge from scientific practice in important ways. Moreover, the contending views in this domain directly intersect with subjects of independent, long-standing metaphysical interest (such as mereology, natural kinds, theories of properties, natural laws, and so on).

One might attempt to explain this lack of attention by pointing out that inasmuch as SAI-ists clearly and loudly prioritize their connection with hard-nosed, ground-level science over airy metaphysics, the case will fall squarely within naturalist territory and will thus be relatively unrevealing about the outer frontiers of a properly naturalized metaphysics. This assumption, I argue, is mistaken. Though generally sympathetic to naturalism—vague and amorphous though it is for me at this point—I do not believe that SAI represents a very clear case for the naturalist. My hope is that this unclarity makes it an interesting and potentially revealing case.

The reason that SAI is an interesting case for considering naturalized metaphysics is, in part, because it is false—or so I have previously argued at length elsewhere (Slater 2013, §§4.1–5.3). Now, this conclusion (if it holds up) tells us little itself about the twin conceptual and methodological questions of what naturalized metaphysics is and how it should be justified and pursued. After all, naturalists need not guarantee that uncontroversial metaphysical characterizations of the relevant science will be easily derived from that science—even for cases apparently well within the naturalist's wheelhouse. (Nor, more obviously still, should they guarantee that all of the science on which metaphysical claims are premised is true.) However, I will argue that the particular *grounds* for taking SAI to be false turn out to be revealing for efforts to naturalize metaphysics, for they are largely metaphysical and a priori in character. Thus, if I am right about SAI, we cannot expect to carry off the project of naturalizing metaphysics simply by always insisting that scientific considerations should trump prior metaphysical convictions; metaphysical convictions may sometimes take priority.

Now, of course, I might be wrong about SAI. My space here to make the case is limited, so I shall have to settle for a mere sketch of the argument. *If* I am wrong, then I offer the case as a schematic depiction of what seems to be a genuinely open possibility—a cautionary fable, if you like. As a consequence, it seems to me that naturalized metaphysics will require a more subtle and open-textured characterization than has yet been offered, and (as a result) will not be very useful as an all-purpose club against a prioristic analytic metaphysics (or against scientific ontologies that adduce reasons falling under that umbrella in their support). We might do better to cleave to the spirit and motivations of naturalized metaphysics, in its full application across the sciences, by assuming a more flexible, pragmatic stance about what counts as good and what bad in the metaphysics of science.

2. THE NATURALIST CLUB

Demarcation projects in philosophy have a checkered history. This fact is not lost on proponents of naturalized metaphysics, as they have faced many of the same difficulties in calibrating a principled and defensible criterion for naturalistically acceptable metaphysics between uninterestingly weak and implausibly strong endpoints. Sometimes, as Penelope Maddy points out, naturalism marks out "little more than a vague science-friendliness"; in order to qualify as *unnaturalistic*, one would have to insist—quite radically—that, for example, "metaphysical intuitions show quantum mechanics to be false" (2007, 1). Even the most imperialistic metaphysicians would abjure that level of science-unfriendliness.

How else might naturalists ratchet up the requirement of "science friendliness"? Another possibility would be to limit the range of acceptable metaphysical projects, for example, those involved in reading off metaphysical theses from scientific theories. Anjan Chakravartty offers some examples: "what ontology of objects and processes is described by the mathematical formalism of theories in fundamental physics [or, we might add, from the special sciences]? Is natural selection a force that acts on some or other biological entity, or is it simply a statistical outcome of causal interactions acting at other levels of description?" (2013, 31). Other naturalists play variations on this basic theme. Ritchie's (2008) "Deflationary Methodological Naturalism," for instance, emphasizes the need for continuity between metaphysics and particular scientific projects, but at the same time cautions against drawing rich philosophical theses (such as physicalism, realism, etc.) from these scientific projects. Morganti offers naturalistic metaphysicians a choice: one should "only do metaphysics as long as it can be immediately applied for the interpretation of science or, alternatively, develop metaphysics independently and then seek application of parts of it" (2013, 22).

On reflection, however, it seems doubtful that mere restrictions to a certain range of projects will ensure a robustly naturalistic orientation. Consider Ladyman and Ross's examples of analytic metaphysicians drawing conclusions about the fundamental mereological atoms that "[lack] any basis in contemporary science" (2007, 19). They might be engaged in a project that overlaps considerably with those pursued by particle physicists. But if the thesis that there are "extended simples," for example, turns out to conflict with quantum mechanics, then, topical cohesion aside, these metaphysicians might feel justified in arguing for the falsity of one of the most empirically successful scientific theories in human history![2] More than simply asking the same questions or operating in the same domain, a full-blooded metaphysical naturalist would also want to insist on a tighter connection between science and metaphysics.

But what should the nature of this connection be? Should metaphysics be relegated to serving as a "handmaiden" to the sciences (cf. Paul 2012), dedicated to merely working out the ontological commitments of our best theories (as Chakravartty's first example illustrates)? Should our metaphysical theories derive exclusively and directly from scientific conclusions? Should methods used in metaphysics be modeled on those used in science? Prominent naturalists have tended to broach such questions by criticizing actual metaphysical practice. For example, Ladyman and Ross criticize the emphasis by analytic metaphysicians on satisfying a prioristic intuitions as a key desiderata in theory construction and confirmation.[3] We can identify two related worries here. First, it is not clear how we might go about vindicating the general reliability of these intuitions. Second, we seem to have some reason to *doubt* their reliability. Ladyman and Ross argue that science has been a consistent source of surprise in precisely the sense that we habitually discover that our intuitive judgments about the world are wrong; we allow science to *show* us this (2007, §1.2.1). Given that preserving and systematizing these intuitions is not—and should not be—emphasized in science, why should things be any different in metaphysics?[4] Pointed questions like this, coupled with cautionary tales of theses proffered as *necessarily true* turning out to be *actually false*, are sometimes interpreted as showing that we ought to have metaphysics either defer to the deliverances of science or adopt its methods.

But there are many forms that such deference can take. It can be absolute, in something like the way that some Logical Positivists envisioned, perhaps leaving no room at all at the intellectual table for metaphysics. Or naturalists can identify *some* scope

[2] Conflicts of this sort can, of course, also arise from a simple lack of awareness rather than any outright hostility to science (or eagerness to see its claims refuted). We might place under this heading the early biological essentialists—apparently science-friendly in disposition—who assumed that biological species would be like chemical species in possessing some "deep-lying" microstructural essence. (How we should think of modern "scientific essentialists" more generally vis-à-vis naturalism, I leave to another occasion.) In any case, a certain level of ignorance can seem like willful disdain given the relative ease with which relevant science might be brought to bear on the metaphysics in question (Ladyman and Ross 2007, 5; Humphreys 2013, 56). A tricky question is how to characterize the proper kind of attention to recent scientific developments.

[3] See also Ladyman's chap. 7 in this volume.

[4] I am merely presenting, not advocating, this line of thought. For two different kinds of critical discussions of Ladyman and Ross on the role of intuition, see Dorr 2010; and Maclaurin and Dyke 2012, §4.

for a suitably reformed metaphysics to contribute to respectable discourse, in which case the question is *how* it should be reformed and constrained. Maclaurin and Dyke contend that the central problem with non-naturalistic metaphysics is that (when it makes ontological claims that extend beyond mere conceptual analysis) it "achieves no practical benefit because [non-naturalistic metaphysical theories like mereology] have no observable consequences" (2012, 301). Correspondingly, they allow that metaphysical debates that *do* have observable consequences will be naturalistically acceptable.[5]

There does seem to be something to the frustration felt by many naturalistically inclined philosophers that many metaphysical debates appear to be empirically inert. But it has proven difficult to formulate and justify a precise principle of deference that avoids both triviality and the unintentional imposition of implausibly strong constraints on science—one might regard *observability* and *practical benefit* as fraught concepts, after all.[6] Ladyman and Ross attempt to give naturalized metaphysics specific content by defending a rather uncompromising series of principles, "referenced to the institutional factors that make science epistemically superior" (2007, 34). One such principle is *the Principle of Naturalistic Closure* (PNC):

> Any new metaphysical claim that is to be taken seriously at time *t* should be motivated by, and only by, the service it would perform, if true, in showing how two or more specific scientific hypotheses, at least one of which is drawn from fundamental physics, jointly explain more than the sum of what is explained by the two hypotheses taken separately. (37)

The PNC distinguishes acceptable from unacceptable metaphysics by placing a different kind of topical constraint on acceptable metaphysics: it should only concern itself with a certain kind of scientific unification project.

I cannot do justice here to Ladyman and Ross's defense of this thesis; in brief, it turns on this unification project being a worthy enterprise that is not credibly left to any particular branch of science (and thus something that metaphysics *could* pursue without gross epistemic negligence). But as Melnyk points out, that this *is* a legitimate philosophical niche for metaphysics to fill does not entail that it is the *only*

---

[5] They offer the debate about the philosophy of time as an example; see Katherine Brading's chap. 1 in this volume for one way in which this might play out.

[6] For critical discussion of Maclaurin and Dyke along these lines, see McLeod and Parsons (2013)—to which Dyke and Maclaurin (2013) respond.

such niche (2013, 82–83). Nor does it seem obvious that fundamental physics ought to be so prioritized.[7]

Rather than seeing naturalistic deference as leading to a constraint on legitimate goals or topics of concern, one might see it instead as a sort of high-level methodology comprising a general encouragement to premise one's philosophizing on developments in relevant sciences (or using their methodological tools) together with a "science-first" policy for conflict resolution: namely, *resolve all apparent conflicts between metaphysics and science by siding exclusively with science*. This sort of "meta-methodology" would presumably only be regarded as a necessary condition of naturalistic metaphysics. Ladyman and Ross hint at something of this flavor in the lead-up to their statement of the PNC:

> Since science just *is* our set of institutional error filters for the job of discovering the objective character of the world—that and no more but also that *and no less*—science respects no domain restrictions and will admit no epistemological rivals (such as natural theology or purely speculative metaphysics). With respect to anything that is a putative fact about the world, scientific institutional processes are absolutely and exclusively authoritative. (2007, 28)[8]

Much more could be said by way of explicating and justifying different approaches to naturalistic metaphysics. But let us take this as a key naturalistic commitment and see how it plays out in the context of the discussions of the metaphysics of species.

3. ANTI-ESSENTIALISM: A NATURALIST SUCCESS STORY?

Even before we had the beginnings of a theory of the *origin* of species, some pre-Darwinian naturalists (in the older sense) recognized the existence of species and sought to understand their properties and relations. From a certain vantage point, such questions might be thought to involve metaphysical speculation from the get-go. What *are* species? Into what ontological category do they fit? The development of evolutionary biology from Darwin's era into the early twentieth century did little to settle this question explicitly, at least in the minds of (even naturalistically inclined) philosophers. Quine, for instance, addresses the question of species' ontological category *en*

---

[7] I do not pretend that these comments add up to a strong argument against the PNC; pursuing this thread further, however, would take us too far afield from my present concerns. For more detailed critical discussion and defense, see the chapters in this volume by Waters and Ladyman.

[8] I take "no epistemological rivals" in a weaker sense than it was perhaps intended to imply the science-first policy of conflict-resolution.

passant in his famous discussion of ontological commitment: "[W]hen we say that some zoological species are cross-fertile, we are committing ourselves to recognizing as entities the several species themselves, *abstract though they are*" (1953, 13; my emphasis). An obvious way of implementing this basic thought, of course, involves treating species as natural kinds. While Quine himself doubted that natural kinds would play any enduring role in mature sciences (1969), his successors were generally inclined to view natural kinds as key targets of scientific investigation (Kornblith 1993). And to many philosophers, species seemed like *paradigmatic* natural kinds in precisely this sense—along with such kinds as gold, water, electrons, and so on. Perhaps inspired by the growing sophistication of molecular biological techniques at the time, Kripke (1980) and Putnam (1975a, 1975b) famously attempted to subsume biological species under the essentialist model, suggesting that, like the physical and chemical paradigms, species would also be characterized by underlying micro-structural essences. More robust arguments for this essentialist interpretation followed over the next few decades (see, e.g., Caplan 1981; Wilkerson 1995; Devitt 2008).

But this nascent metaphysics of species—that they are natural kinds characterized by essences—did not remain in favor for long. Several influential evolutionary biologists had already been campaigning against treating species as abstract types in either a Platonic or Aristotelian conception (see, e.g., Mayr 1959, 28–29; Ghiselin 1966, 1974). Biologically sophisticated philosophers gradually took up the thread and pronounced the abject failure of natural kind essentialism as applied to biological species. Elliott Sober wrote in a well-known article that "[e]ssentialism about species is today a dead issue, not because there is no conceivable way to defend it, but because the way in which it was defended by biologists was thoroughly discredited" (1980, 353), and because it was a poor explanation of the patterns of variation we observe. Others focused less on the manner of essentialism's defense than on the empirical falsity of its core commitments: that for each species, there is some underlying microstructural (presumably genetic) property shared by all and only its members that makes them the sorts of things they are. Increasing understanding of the nature of development and the extent of intraspecific genetic variation rendered these claims significantly less plausible.

So we have here what seems to be a pretty obvious victory for the naturalist. A metaphysics of species is proposed and subsequently falsified by empirical investigation. Of course, the story is a bit more complicated than this brief gloss lets on—but perhaps only a bit. For one, the historical narrative offered by Mayr (and taken up by many philosophers of biology) concerning the development of essentialism has been questioned (Winsor 2003, 2006). For two, Sober's pronouncement of essentialism's death might have been, as they say, greatly (or anyway *somewhat*) exaggerated (Devitt 2008). Essentialism about biological taxa has persisted since the

1980s in two forms that we might call *historical* and (tendentiously) *bullet-biting*. The former involved abandoning one of the standard tenets of essentialism: that a natural kind's essence must be an intrinsic property. Perhaps species kinds could be seen as defined by *historical essences*—for example, by their position on the tree of life in relation to other taxa (Griffiths 1999; LaPorte 2004; Williams 2011).

Let us consider the bullet-biting strategy for defending essentialism first. One might pursue it in two different ways. First, one could propose simply revising the boundaries of the relevant kinds so that they could be characterized by intrinsic essences (e.g., Wilkerson 1995, 132). Second, one might leave recognized taxa alone and argue that they just *must* have essences, even if so far unidentified and maybe also even if there is specific reason for *doubting* that there are such properties. Devitt, for instance, argues from the existence of lawlike generalizations about species (e.g., "All Indian rhinos have a single horn") that these generalizations must be explained by intrinsic biological essences. Though this gambit leaves biological classification more or less unmolested, it does seem to entail that key pieces of biological *practice*—the way in which systematists have gone about describing species and their relations—are seriously misguided. Devitt's argument would, at the very least, imply that biologists have been systematically ignoring a key explanatory burden in biology (cf. Lewens 2012b, 752). This alone might suggest to the naturalistically inclined metaphysician that something must be amiss in Devitt's arguments (or in the other bullet-biting strategies).[9] Contemporary biological practice, like most cases of scientific practice, is motivated by its record of bearing fruit (or promise for doing so) on matters deemed significant by the community pursuing it. As such, it should not be abandoned (or significantly modified) lightly.[10]

Contrast the historical essence strategy mentioned above. In that case, it was (interpreted as) standard practice of evolutionary biologists to treat species as historically defined kinds so that, as Griffiths puts it, "nothing that does not share the historical origin of [a given] kind can be a member of [that] kind" (1999, 219). Though this approach does less violence to large portions of biological practice—effectively leaving alone cladistically inclined classification schemes—it does apparently foreclose on some apparently live options about how to characterize species. Suppose one was interested in pursuing species concepts based on shared gene clusters (Mallet 1995), interbreeding (Mayr 1963; Coyne and Orr 2004), or phenetic similarity

[9] Though Devitt offers some brief comments concerning the use of genome sequencing projects in revealing the intrinsic essences of various taxa—suggesting that he *is* concerned with biological practice—one does not typically observe biologists discussing these projects in terms of the discovery of species' essence.

[10] Devitt's arguments may of course fail on their own terms, as I have argued they do (2013, §3.3); see also Barker 2010.

(Sokal and Sneath 1961; Lewens 2012a). The restriction on sharing a historical origin Griffiths posits would have to be grafted onto these species concepts artificially and without internal scientific motivation. This is analogous to the awkward situation Callender discusses concerning the "strange new laws" some metaphysics are forced to postulate in order to accommodate their metaphysical views about "extended simples" to findings in theoretical physics (2011, 39).

So a comparatively mild naturalistic rebuke seems appropriate here as well. To those with even slightly heterodox or pluralist approaches on the species problem, historical essentialism (biological motivation notwithstanding) overreaches. Of course, this rebuke may not move those antecedently committed to classification strategies that already incorporate Griffiths' stricture. This raises a question about what degree of scientific ecumenicalism a naturalized metaphysics should allow. But this is a question I must leave aside here. For now, let us suppose that, a few complications aside, the anti-essentialist consensus in the philosophy of biology is a case of due attention to the relevant science foreclosing on otherwise appealing metaphysical approaches. I want to now turn toward two other cases—one positive, one negative—that do not transfer quite as cleanly. We start with the positive: the claim that species are individual objects with organisms as their parts (rather than natural kinds). We will then consider a criticism of an alternative conception of species as natural kinds.

4. INDIVIDUALISM

The thesis that species are individuals (SAI) was originally motivated in large part by the so-called Species Problem. As standardly described, the problem is how to formulate a satisfying response to the long-standing dispute among biologists about how to characterize species (Wilkins 2009). One can interpret this challenge in two different ways: first, there is the challenge of actually *resolving* the ground-level dispute by, say, showing conclusively (or satisfactorily) that a particular conception of species was correct; second, there is the challenge of squaring the prevalent sense that species are in some sense real, objective features of the world with the fact that even after decades (if not centuries), biologists cannot agree on how they should be defined. If species are real, what sort of features of the world are they?

SAI was originally proposed by the evolutionary biologist Michael Ghiselin as a solution to the Species Problem construed in this second sense.[11] Here is the basic thought: rather than thinking of species as natural kinds characterized by

[11] Indeed, the 1974 paper in which SAI received its fullest initial presentation was titled "A Radical Solution to the Species Problem."

intrinsic essences, we should regard them as spatiotemporally extended concrete particulars—persisting composite objects made up of their organisms. Unlike paradigmatic objects such as rocks or tables, these species objects would typically be scattered *at a time*, but they would nevertheless be connected *over* time by the sorts of lineage-forming relations (such as interbreeding and gene-exchange) that biologists often employ for demarcating species in the first place. The objectivity of the parthood relation's obtaining would secure the objectivity of species against the threat posed by the species problem. Whatever disagreement there was about how to conceptualize species, there would be some fact of the matter about how organisms were united into larger, persisting individuals, rendering (so the thought went) some concepts objectively better than others at mirroring this structure.

But why accept SAI? That it *would* solve a problem (if true) does not, by itself, show that it *is* true. Without surveying all of the relevant arguments (see Slater [2013, §4.2] for more), it is worth mentioning a representative few. In the first place, concerns about essentialist models of natural kinds (alternatives to which were not clearly seen when SAI was first proposed) loomed large. In a two-category ontology—of universals and particulars (or abstracta and concreta), say—a simple eliminative inference suggested that, if they are not abstract types, species must be concrete particulars. This move cohered nicely with Ernst Mayr's campaign to embrace "population thinking" as an alternative to "typological thinking." On this line of thought, species are certain sorts of groups of populations (or better: *metapopulations*), and these are not like *properties* that things can have; they are *things* with properties themselves, with a certain spatiotemporal extent—in other words: they are individuals.

A second, and related, style of argument focused on pressing for an analogy between the way biologists think of organisms and how they think of species—for example, in how much disruption or change they can endure or in the level of "cohesion" among their parts (Hull 1978, 347; 1989, 84; 1999).

A third argument strategy highlighted the explanatory power of the SAI thesis. Why are species spatiotemporally restricted (Ghiselin 1987, 128)? Why are there no natural laws about particular biological species (Hull 1977, 150)? Why are biologists hesitant to speak of organisms as "instances" of species—of a particular organism exemplifying the property of "tigerhood," say (Ghiselin 1974, 536)? All of these questions, the argument goes, would receive satisfying and unifying answers on the thesis that species are individuals. Thus we should tentatively conclude that SAI is probably right.

A certain naturalistic orientation is clear in each of these broad styles of argument. The first is premised on empirical considerations coming directly out of biology. It is an extension of the apparently successful naturalist critique of essentialism (setting the aforementioned complexities aside). The second stems from the attitudes of biologists in how they think about and refer to species. Responding to philosophers' objections to the argument-by-analogy strategy (e.g., Kitcher 1989), Richards notes,

> Those who work in the biological sciences typically don't see such a distinctive and important disanalogy between individual organisms and individual species taxa. There are, I believe, several reasons based on familiarity with biodiversity, tradition and disciplinary practices. . . . A full consideration of biodiversity reveals the bias in philosophers' commonsense notions of individuals, and its focus on vertebrates and humans. (2010, 163–64; see also de Queiroz 1999, 67)

So perhaps biologists' training and experience in their science disabuses them of the need to see individual objects as spatially cohesive—rather like experience in physics disabuses us from naïve conception of objects as solid and continuous.[12] The final argument strategy is transparently naturalistic, using explanatory patterns of inductive inference familiar from the sciences and drawing on purported facts that while familiar to biologists might seem surprising to the uninitiated layperson. After all, this gap between expert and lay beliefs about species could be cited as a main factor behind philosophers' essentialist foibles when it came to species.

Relatedly, some SAI advocates have argued in a Quinean spirit that the way that biologists refer to species strongly legislates for the individualistic interpretation. Unlike the lay who might think of, for example, *being a tiger* as a property that things can possess, these SAI advocates will point out that the practice of demarcating species historically indicates that "tiger" functions instead as a singular term. Here is Coleman and Wiley's version of the argument:

> Biological theory is replete with generalizations that seem to be about particular things called "species." The generalizations "There are species" and "Species are variable" appear to be statements that are truly or falsely said about species and not about the organisms that comprise a species. . . . Thus, one way to interpret discourse about species is to understand at least some of the expressions used to talk about particular species taxa as genuine singular terms (i.e., as terms referring to particular things rather than designating kinds of things). This manner of interpreting talk about species comprises what we call an "objectual account of species." (2001, 499–500)

---

[12] Again, I am not granting these points here. My present focus is descriptive and interpretive. Barker and Wilson (2010) have argued compellingly that the kind of cohesion enjoyed by the members of a species is not especially compelling as an integrating force.

This argument from ontological commitment would of course be stronger if it could be shown that biological discourse could *only* be captured by treating species names as singular terms—something that Coleman and Wiley attempt later in their article.

Even if none of these arguments are conclusive, it would not be unreasonable to feel swayed by their collective weight. For they each appear to be grounded, in one way or another, in our best biological science. Add to this the consensus among philosophers of biology and biologists attending to the matter, and it can start even to seem *foolish* to suppose that one could argue against SAI from a metaphysical perspective—particularly if one felt any naturalist sympathies. Would this not be tantamount to a non-naturalistic stance? I will say *no*; but vindicating this answer requires articulating a different vision of naturalized metaphysics. While I will not be in a position of completing this task in any detail in this chapter, I want to set off in this direction by considering a way in which (apparently) purely metaphysical considerations tell against this popular naturalistic metaphysics of species and asking how we should think about this debate from a naturalistic perspective.

The considerations in question emerge from attention to the metaphysics of vagueness and indeterminacy over the past few decades, specifically as they concern the possibility of indeterminate parthood, identity, and existence. A good starting point is Gareth Evans's much discussed (1978) paper purporting to demonstrate the incoherence of ontic indeterminate identity—that is, the incoherence of identity statements that are indeterminate in truth-value *not* due to any semantic defect or imperfect knowledge. Like many later commentators, Evans took his argument as addressing the coherence of the idea that "the world itself might *be* vague. Rather than vagueness being a deficiency in our mode of describing the world, it would then be a necessary feature of any true description of it" (208). The argument takes the form of a *reductio.* Suppose that some thing(s) *a* and *b* are indeterminately identical (and that this is a metaphysical, not a semantic fact due to referential ambiguity). Another way of putting this is to say that individual *a* has the property of *being indeterminately identical to individual b*. However, *b* does not have that property: it is definitely itself! So there is some property—*being indeterminately identical to b*—that *a* has but *b* lacks. Since, by Leibniz's Law, something cannot have different properties than it has, it must be that *a* and *b* are in fact distinct; and if they are provably distinct, they are not indeterminately identical after all—contra our supposition.[13]

Evans's result has been extended to contest the possibility of indeterminately *existing* objects. In Ted Sider's argument, we begin again from a supposition

[13] Lots more could be—and has been—said about the Evans argument and the issues surrounding ontic vagueness; for helpful discussion, I would see in particular Lewis 1988; van Inwagen 1988; Heck 1998; Moore 2008; and Barnes 2010.

(for *reductio*) that there are indeterminately existing objects. From this it follows that there are numerical sentences expressing how many objects there are—sentences of the form "$\exists x\, \exists y\, \exists z \ldots (x \neq y \,\&\, x \neq z \,\&\, y \neq z \ldots)$"—that are indeterminate in truth-value. But there is nowhere for that indeterminacy to lie, as the vocabulary of such sentences contains only sentential connectives, unrestricted quantifiers, and identity, none of which can admit of indeterminacy. So the sentences are not indeterminate in truth-value after all; *reductio* achieved (Sider 2001, 125–30).[14]

What does the prohibition against ontic indeterminate existence and identity have to do with SAI? The problem is that biological discourse concerning species concepts exhibits little interest in obeying the strictures against ontic indeterminate existence. Worse, it displays tendencies in the opposite direction. Every species concept of which I am aware allows some indeterminacy in its application.[15] Yet this discourse is what tells us how organisms are united into species—including whether some organisms comprise one species or multiple species. Thus if we interpret SAI as a metaphysically robust account of what species are—that they are *real, left-alone objects*—our best biology foists upon us an apparently intolerable conclusion: that some objects only indeterminately exist (or are indeterminately each other).[16]

Call this the Indeterminacy Problem for SAI. I think that it provides us with strong reason for rejecting SAI, despite that view's strong naturalistic credentials. Now of course there are a number of ways in which an SAI sympathizer might respond. Let us set aside responses that significantly weaken the metaphysical import of the thesis—say, by construing it as a sort of *façon de parler* or a semantic thesis of some

---

[14] As with Evans's argument, Sider's has not been without its critics—to whom Sider has responded (see Koslicki 2003; Sider 2003; Liebesman and Eklund 2007; Sider 2009; Barnes 2010). And while there is certainly room to press further, my sense is that its conclusion is extremely plausible. Even Barnes's (2010) critical discussion of Sider (2003) allows that "[i]ndeterminate existence is deeply problematic when interpreted *de re*. The claim that there is some thing such that it's indeterminate whether that thing exists is hard, if not impossible, to make sense of" (960). David Lewis expressed this same thought when he asked (rhetorically) of a putatively indeterminately existing object, "What is this thing such that it sort of is so, and sort of isn't, that there is any such thing?" (1986, 212).

[15] I document this in some detail in Slater (2013, §4.3.2), but to get the flavor, we can consider a popular species concept like Mayr's "Biological Species Concept" (BSC), according to which a species is a group of "interbreeding natural populations that are reproductively isolated from other such groups" (Mayr 1963, 17). However, reproductive isolation—at least, as it is used in this species concept—is not generally a sharp notion. Whether two populations count as members of one species or two sometimes depends on the degree to which their members can exchange genes (Sterelny and Maclaurin 2008, 28). One might argue that it is possible to sharpen this criterion and avoid the threat of indeterminacy. But while there are certainly *some* pairs of organisms for which such isolation is very much an all-or-nothing matter, a general insistence on sharpness leads to some implausible consequences that biologists are not at all eager to embrace. Indeed, many explicitly acknowledge a certain level of vagueness in their account (see, e.g., van Valen 1976, 234).

[16] One might hope for an epistemic or semantic resolution of the Indeterminacy Problem. Alternatively, one might attempt to downplay the significance of ontic indeterminacy—for example, by suggesting that it is in fact *everywhere*. I discuss why I believe that these strategies will fail in Slater (2013, 92–95).

kind. These may be worth considering, but they are of little interest in the context of a discussion of naturalized *metaphysics*. One might attempt to contain the problem for SAI without engaging directly with the metaphysics of indeterminacy by reinterpreting the SAI thesis in a non-mereological way (Haber 2013, 2015). At best, I think this postpones answering key questions: How are *biological parthood* and existence (or *biological* existence?) conceptually linked? What is the metaphysics of *biological parthood* such that it evades the indeterminacy problem? Does it do so, for example, by taking a deflationary approach to the metaphysics of species? If not, then so long as SAI involves ontological claims about what exists and what is identical with what, it is not clear to me how the precise formulation of a theory of biological parthood will be able to address this problem (Slater 2013, 89).

A more direct way of responding, of course, engages directly with the Evans/Sider arguments. Might a naturalist simply "recalcitrate" (as Quine put it) and take the existence of indeterminately existing (or indeterminately identical) species for granted and use it as evidence that the Evans/Sider arguments are unsound? Of course; but doing so is unsatisfying. Where specifically does the argument go wrong (cf. Dorr 2010)? Suppose that the SAI-naturalist replies as follows: *I do not need to tell you where these arguments go wrong. I have the falsity of the conclusion—that indeterminately-existing things are possible—directly from some very well-confirmed science (evolutionary biology and systematics). All you have is your a priori argument that this is impossible. But since such empirical considerations trump your* a priori *conclusions, we can be confident that these arguments must go wrong somewhere.*

This sort of response will not do as is, however. For one, it is obviously false that the empirical should *always* trump the a priori. Imagine an investigator coming to a conclusion at odds with some piece of pure mathematics; she would scarcely consider thinking "well, so much for arithmetic!" Rather, she would double-check her work and look for a mistaken assumption or botched procedure, explaining to colleagues that the numbers just did not add up.[17] Now, one might object at this point that mathematics constitutes a principled exception to the above ranking of the empirical above the a priori and that our imagined naturalist's reply should be interpreted as committed to the thesis that the empirical always trumps a priori *metaphysics*. But having made one exception, it seems reasonable to wonder how we might recognize

[17] This is not, of course, to say that the reverse tie-breaking principle should be employed. As noted above, we have seen purportedly a priori axioms of Euclidean Geometry overturned by results from physics. And lest we chalk that episode up as a curious historical anomaly to be reclassified somehow, note that more recent thinkers have felt pressed, if not moved, to reconsider other a priori claims on the basis of empirical discoveries. For example, Quine (1981) entertained the possibility that certain interpretations of quantum mechanics might even compel us to drop such logically sacrosanct principles as Bivalence.

others. Moreover, one might doubt that relative epistemic weight is exhausted by the epistemic modality by which we come to believe something.

Relatedly, and more to our present concerns, E. J. Lowe (1994) argued that certain physical scenarios involving quantum entanglement reveal the falsity of Evans's conclusion about the impossibility of ontic vague identity. Without descending too far into the detail, he considers a scenario in which an electron is absorbed by an atom and another electron is later emitted. Is the emitted electron the same as the absorbed electron? Lowe claims that there is no saying one way or the other (due to quantum entanglement); but not because we do not know enough or because our names are fuzzy. No, the indeterminacy is ontic. So there we go: ontic indeterminate identity, contra Evans, is possible. And this is shown by one of the best-confirmed scientific theories in the history of science.

The case is not so simple, however (for discussion, see Noonan 1995; Hawley 1998; French and Krause 2003). Nor is the comparison obviously helpful for supporters of SAI. For one, and to Lowe's credit, he did not simply invert a modus ponens into a modus tollens; he offered a diagnosis of the Evans argument. More significantly, as the critical discussion of the paper made clear, a good deal of interpretation of the robust experimental and theoretical results of quantum mechanics is necessary in order for Lowe's case to do the job it was meant to do. One has to do some metaphysics of science in order for the experimental results to bear on further metaphysical matters. This appears to be a case in Chakravartty's general point that "even on the most metaphysically austere, contemporary conception of the sciences . . . there is a metaphysical dimension to [scientific] inquiry" (2010, 70).

This illustrates a second reason why we should find the recalcitrating response to the Indeterminacy Problem unsatisfying: the conflict is not *simply* between a metaphysical thesis about ontic indeterminate existence and a robust empirical result; in fact, both theses have metaphysical content. Suppose that the SAI-advocate argues on naturalistic grounds that the closer proximity of their metaphysical thesis to empirical science breaks any imagined epistemic tie: on the one hand, we have a metaphysical thesis that is continuous with current science; on the other, we have speculative a priori metaphysics at great remove from empirical science. We should not be happy with this characterization of the dialectical situation, however. In the first place, the asymmetric labeling of these sources as "speculative" begs important questions; and anyway, scientific conclusions and observations can often themselves be highly speculative. But set this concern aside and grant that the relevant empirical results in biology are as epistemically secure as you like. This is just one part of the story. While there is a certain temptation to gauge the relative epistemic priority of the competing theses on the degree to which they involve metaphysical or empirical premises, the contest should also involve the strength and character of the *inferences* from these premises.

Here matters become shakier and dependent on further metaphysical assumptions, especially when the inferences are eliminative (e.g., when one argues that species must be individuals because they are not natural kinds), as these inferences assume that we have seen clearly all of the relevant alternatives—something that history shows is often unreasonable to assume (Stanford 2006).[18] Consider Haber's claim that "species are spatiotemporally located, concrete entities, with a beginning and an end, and have parts, not members, is a function of how lineages are generated, and of the facts of reproduction" (2015, 15). But it is not at all obvious how the biology of reproduction entails these further characterizations without the intervention of further metaphysical theories. Biology can tell us how lineages in fact form, but it does not without significant additional interpretation tell us how they *must* form or how we should conceptualize them metaphysically.[19]

5. HOMEOSTATIC PROPERTY CLUSTER ANALYSES

Let us examine a final case in which naturalistic grounds for advocating a particular thesis on the ontological status of species do not clearly carry the day. In the case of SAI, we considered concerns over a positive metaphysical account despite its being naturalistically motivated. This may be disappointing for naturalists, but perhaps it simply represents an overreach. It thus may be instructive to consider a more modest critical argument against a particular metaphysical perspective on species that proceeds on naturalistic grounds but which also (I think) fails.

The case involves Boyd's Homeostatic Property Cluster (HPC) account of natural kinds (Boyd 1999a), an account that many philosophers of biology argue evades the criticisms levied against essentialistic conceptions of natural kinds (see, e.g., Rieppel 2007; Wilson, Barker, and Brigandt 2007). Species might be natural kinds, on this view, in virtue of sharing a cluster of properties, perhaps imperfectly, whose broad stability/sociability is maintained via the operation of certain homeostatic mechanisms. In my view, the HPC account faces a number of serious theoretical issues and problems for its application to species taxa; I am not an advocate (for criticism, see, e.g., Häggqvist 2005; Reydon 2009; Slater 2013, §6.2.2; 2015). But it has recently come under some criticism that I believe represents naturalist-overreach.

A chief critic has been Marc Ereshefsky, who has argued in several papers that HPC theory offends from prevailing views in biological systematics. Naturalistic concerns are clearly front and center in his argument; he writes (with

[18] For some suggestions of previously unconceived alternative on this front, see Slater (2013, chap. 5).

[19] I am indebted to Celso Neto for comments that helped clarify my arguments in this section.

Mohan Matthen): "The HPC account of biological taxa should be informed by, or at least consistent with, biological systematics. But that does not seem to be the case: the historical essentialism of HPC theory is not historical enough for contemporary biological systematics" (2005, 17). Why not? The answer stems from some critical comments Boyd levies against constraints certain systematists insist upon but that Boyd finds to be unmotivated (1999a, 180–82; 2010)—in particular, he allows that HPC kinds might be para- or polyphyletic. Here Ereshefsky objects, "This should not rest easily with those who believe that [species or] higher taxa must be monophyletic" (2007, 296).[20]

But the issue here is a bit more subtle than Ereshefsky allows. We can agree to the insistence that HPC theory ought to be informed by and consistent with empirical results in biology; but it is not clear that it must be *just as demanding* as a particular school of systematics that not all biologists happen to advocate. This would be a very strong and implausible interpretation of consistency. Does consistency with biology require that biologists be disposed to adopt (or at least not reject) the theory in question? That is not obvious either. I argued above that SAI's popularity among biologists was not particularly probative; likewise, the fact that some biologists may be expected to reject HPC theory does not, in and of itself, show that it is inconsistent with biology. After all, biologists—even those with a phylogenetic orientation—disagree with each other about how to organize taxa (de Queiroz 1999; Wheeler and Meier 2000). Consistency with biological systematics will remain elusive as long as extant theories of biological systematics are not consistent with each other.

But set this biological conflict to one side. Suppose (what is radically contrary to fact), that biologists *were* united in advocating monophyly as a constraint on how species and higher taxa ought to be individuated. Would this not reinstate the force of the naturalist criticism of HPC? Perhaps it would make the case stronger, but again I do not believe that HPCers ought to feel forced to capitulate. For just as we are entitled to differentiate between the epistemic worth of particular empirical theses, we should take seriously the potential for acknowledging epistemic differences between such theses and prevailing classificatory practices. While it is tempting to place all aspects of scientific practice onto the same epistemic pedestal as its acknowledged fruit, we should recognize the possibility that such practices do not contribute

[20] A quick and dirty primer on some of these terms: taxa are monophyletic if they consist of all and only the descendants of a common ancestor; advocates of the monophyly requirement are sometimes called "cladists," though (confusingly) this term is also sometimes used interchangeably with more general "phylogenetic" approaches to systematics that see historical approaches to taxonomy as fundamental. Different phylogenetic approaches might recognize paraphyletic taxa, which contain *only* descendants of a common ancestor, but not *all* of them (it might leave a branch off) or polyphyletic taxa, dropping the "*only*" requirement as well: they may comprise some descendants of multiple lineages.

equally to the epistemic worth of science and so do not clearly merit trumping force when apparently in conflict with metaphysical theses. Similar considerations can be applied to naturalistic arguments stemming from the ways in which a particular community of scientists talk. If there are other possible ways of classifying, investigating, and talking that would lead scientists to comparative success, we ought to be cautious about reading off conclusions from these considerations.

I will return to some consequences of this stance in the final section. But before we move away from our final case study, it is worth noting two further contrasts between the (in my view) successful anti-essentialist case and this unsuccessful anti-HPC case. In the former case, essentialists were making a claim with content that conflicted with a well-established empirical result in biology—for example, that species taxa share a common "genetic structure" that makes them the species they are. That seems not to be true (nor should we expect it to be true, given other things we know about evolution, development, and the sources of variation). In the latter case, it is less clear that the conflict between HPC and cladists has the same empirical character. Cladists may say "biological taxa are monophyletic," but it is not evident that this is an empirical claim in the same way that "biological taxa are genetically heterogeneous" as much as it is a *normative* claim about how we ought to group organisms together—that is, "biological taxa should be grouped according to monophyly." Such a claim is not, as it were, read off from the data. Granted: it is an oversimplification to suppose any scientific claim can be so read, or regarded as "true on inspection," but arguably there is a qualitative difference here.[21]

There is more to say here, of course, about classificatory norms—how they interact with empirical content of scientific theories, how they come to be, how they are embedded in scientific practice more generally. My discussion has so far remained rather general (and so it must remain in this context).[22] It is premised on an undefended but I think plausible claim: that classificatory practice, like any component of scientific practice, results from a complex interaction between contingent normative commitments about values, aims, and significance-ascriptions and empirical facts (which are in turn partially conditioned by those norms); such complex interactions will involve trade-offs and thus tend to demand a pluralistic outlook in those cases. But there is no similar call to be pluralistic about claims involving the degree of genetic homogeneity we find in certain kinds of biological taxa (so long as one fixes on a particular understanding of the taxa involved).

---

[21] Note also that even if it turned out that species are substantially artifacts of our classificatory whims, the empirical content of these claims would still differ.

[22] In other work (Slater, n.d.), I have argued that much of classificatory practice is guided (and not determined) by various classificatory norms that, while responsive to empirical facts, are not fixed by them.

The second contrast with the anti-essentialism case is the way in which essentialism and views like HPC may be applied to biological taxa. Devitt's "Intrinsic Biological Essentialism," for example, is the thesis that "Linnaean taxa have essences that are, at least partly, intrinsic underlying properties" (2008, 346). This could be interpreted either as an empirical claim about the genetic structure of already identified (and identifiable) taxa or as a normative constraint on how taxa should be identified (or, if need be, revised). On either interpretation, there is a conflict with our best science—with what it takes to be true or how it goes about systematics. HPC, on the other hand, drops the former empirical claim and offers a far more flexible framework for recognizing "natural kind phenomena." In my view, treating species as HPC kinds need not be taken as a constraint upon scientific practice (cf. Boyd 1999b, 162–63). Ereshefsky's criticism is that HPC takes the wrong orientation to identifying taxa. While he allows that "there may be many HPC kinds in the world," he seems to deny that *any* biological taxa are among them: "HPC's emphasis on similarity is at odds with phylogenetic approaches to taxonomy. HPC kinds are fundamentally similarity classes, whereas taxa are fundamentally genealogical entities" (2007, 296). But this ignores the possibility that "genealogical entities" may also have members (or parts, if you insist) that share certain clusters of properties in common due to the activity of certain homeostatic mechanisms. The fact that being an HPC kind involves no genealogical qualifications clearly does not mean that things that *do* meet such qualifications cannot be HPC kinds; a metaphysics of species need not be the sole source of factual claims about species—it had *better* not be if we are to remain in generally naturalistic territory.[23]

6. CONCLUSIONS

Having now considered these three cases, what shall we say about the plausibility of specific naturalistic constraints on metaphysics? First, I believe that the cases illustrate (what seems independently plausible) that not every apparent demand that issues from science should be accorded equal weight. Not all scientific claims are on an epistemic par; nor are all metaphysical claims. We cannot, then, assume that entire disciplines can be neatly stacked in terms of their epistemic superiority. Perhaps some claims that appear to be metaphysical in nature—assuming that this status can be defined independently of their evidential grounds—are in a position of trumping claims extending from the natural sciences. At the level of conflict

[23] To be fair to Ereshefsky, it sometimes seems that Boyd is using HPC to criticize the monophyly requirement imposed by some schools of systematics. I do not have the space to evaluate this framing of the dispute; the issues here are complex.

between particular claims, science *might well* admit of epistemological rivals even if it is unrivaled as an intellectual/epistemic corpus. We cannot reasonably prioritize the scientific over the metaphysical any more than we may reasonably prioritize beliefs formed by, say, vision over those formed via testimony as justificatorily superior.

Second, in evaluating and resolving apparent conflicts, we need to take into account both the relative epistemic standing of the competing claims and the role that potentially silent auxiliary metaphysical theses might play in generating that competition. Third, and relatedly, we should be cognizant of a problem of unconceived alternatives for scientific metaphysics and draw inferences about the correct metaphysics of species (or what have you) cautiously.

As these conclusions function as partial checks on naturalistic constraint, the story so far is largely negative. Is there anything we can say at this point on the positive side about how we ought to naturalize metaphysics? Are there principles for resolving the sorts of apparent conflicts we have examined between scientific and metaphysical claims? How, in general, can we give specific, defensible, and non-vacuous content to the concept of naturalized metaphysics? I am not confident that we can. This is not to deny that there are many cases of metaphysical overreach where naturalistic ideals are flagrantly ignored. But we can often diagnose such cases—as illustrated by essentialism about species—as metaphysics failing to take into account well-established facts that bear on the theses in question. I take it, however, that such input can come from any source. Science is generally good at producing claims that deserve to be taken seriously. Many of these obviously bear on philosophical claims. Thus they should be sought out and acknowledged by philosophers and taken into account as appropriate. It is this last imperative that seems to me difficult to assign precise content. My suggestion that we distinguish between scientific practice, norms of classification (or other aspects of scientific practice), and empirical findings, theories, proposals, and working hypotheses (and distinguish further within these categories according to their epistemic standing) may be part of the story, but I do not assume that we can map very precise methodological constraints onto such divisions.[24]

This story might change somewhat depending on how we understand the project of scientific metaphysics (or metaphysics generally). I argued in the previous

---

24 I read McLeod and Parsons as making a similar claim when, in the finale of their critical response to Maclaurin and Dyke's previously discussed criterion for naturalistic acceptability (2012), they write that "it's not possible to determine which theories are [naturalistically acceptable] just by understanding what those theories say. To determine whether a theory is naturalistic, we have to do some philosophy (and possibly some science too)" (2013, 177–78).

section that naturalistic constraint should be at best hesitant in contexts in which pluralism may be expected to hold sway (say, in cases where methodological norms or classificatory practice was largely contingent). Even if a particular piece of scientific practice has developed in a uniform way, if it might have legitimately developed differently (with different metaphysical consequences), then the naturalist cannot reasonably insist on grounds of scientific consensus that the corresponding metaphysics must be this or that other way. But suppose we adopted an approach to metaphysics like the one Peter Godfrey-Smith describes, according to which "the theoretical constructs developed in systematic metaphysics are best seen as models. Metaphysical system-building is model building" (2006, 6). On this view, concerns about pluralism and contingency fall away. A consensus classificatory practice might be as contingent as you like; we can still model it and discuss its ontology in this deflationary sense without any pretense that we are mirroring the fundamental structure of nature.

Your mileage may vary; maybe this does not seem like doing *metaphysics*. Fair enough. Yet perhaps we can maintain something of the spirit of Godfrey-Smith's conception while moving in the direction of (traditional) metaphysics naturalized. Here I am reminded of some remarks of Mark Johnston's at the outset of an important essay on how we might formulate a defensible, non-verificationist Pragmatism; he writes,

> Let us say that metaphysics in the pejorative sense is a confused conception of what legitimates our practices; confused because metaphysics in this sense is a series of pictures of the world as containing various independent demands for our practices, when the only real legitimation of those practices consists in showing their worthiness to survive on the testing ground of everyday life. Then metaphysics is not just a technical discourse within philosophy to which, since Kant, a technical apparatus of philosophical criticism has been opposed. It is endemic to our culture. So defined, metaphysics is the proper object of that practical criticism which asks whether the apparently legitimating stories which help sustain our practices really do legitimate, and whether the real explanations of our practices allow us to justify them. There then ought to be a critical philosophy which not only corrals the developed manifestations of metaphysics within philosophy but also serves the ends of practical criticism. Such a critical philosophy would be the content of anything that deserved the name of a progressive Pragmatism. (1993, 85)

Some versions of naturalism can be characterized as strong reactions against Johnston's "metaphysics in the pejorative sense"; I have argued that such naturalists err in asserting too strong an asymmetry in the opposite direction, "corralling" metaphysics too aggressively. I do not claim that naturalized metaphysics should be seen as an outcropping of pragmatism. Rather I see a progressively naturalized metaphysics as a regulative ideal in which competing demands of practice, norms, features of disciplinary discourse (both scientific and metaphysical), and our best theories (both scientific and metaphysical) are weighed against one another in order to develop the best picture of the world that we can manage. This pragmatic stance seems superior to a discipline-centered approach in its ability to avoid the problems discussed above while making flexible use of our epistemic resources. If more can be said in this direction, we will just have to evaluate such proposals on their merits.

ACKNOWLEDGMENTS

For helpful comments on previous drafts and presentations of this material, I am indebted in particular to Matt Barker, Richard Boyd, Olivier Lemeire, Alan Love, Celso Neto, Jay Odenbaugh, William Wimsatt, Zanja Yudell, and audiences at the 2013 Pacific, 2016 Central APAs, and at Lewis & Clark College.

REFERENCES

Barker, Matthew J. 2010. "Specious Intrinsicalism." *Philosophy of Science* 77: 73–91.

Barker, Matthew J., and Robert A. Wilson. 2010. "Cohesion, Gene Flow, and the Nature of Species." *Journal of Philosophy* 107 (2): 61–79.

Barnes, Elizabeth. 2010. "Arguments against Metaphysical Indeterminacy and Vagueness." *Philosophy Compass* 5 (11): 953–64.

Boyd, Richard. 1999a. "Homeostasis, Species, and Higher Taxa." In *Species: New Interdisciplinary Essays*, edited by Robert A. Wilson, 141–185. Cambridge, MA: MIT Press.

Boyd, Richard. 1999b. "Kinds, Complexity and Multiple Realization." *Philosophical Studies* 95: 67–98.

Boyd, Richard. 2010. "Homeostasis, Higher Taxa, and Monophyly." *Philosophy of Science* 77 (5): 686–701.

Briggs, D., and S. M. Walters. 1997. *Plant Variation and Evolution*. 3rd ed. Cambridge: Cambridge University Press.

Callender, Craig. 2011. "Philosophy of Science and Metaphysics." In *The Continuum Companion to the Philosophy of Science*, edited by Steven French and Juha Saatsi, 33–54. London: Continuum International Publishing Group.

Caplan, Arthur L. 1981. "Back to Class: A Note on the Ontology of Species." *Philosophy of Science* 48: 130–40.

Chakravartty, Anjan. 2010. "Metaphysics Between the Sciences and Philosophies of Science." In *New Waves in Philosophy of Science*, edited by P. D. Magnus and Jacob Busch, 59–77. Basingstoke: Palgrave Macmillan.

Chakravartty, Anjan. 2013. "On the Prospects of Naturalized Metaphysics." In *Scientific Metaphysics*, edited by Don Ross, James Ladyman, and Harold Kincaid, 27–50. Oxford: Oxford University Press.

Coleman, Keith A., and E. O. Wiley. 2001. "On Species Individualism: A New Defense of the Species-as-Individuals Hypothesis." *Philosophy of Science* 68: 498–517.

Coyne, Jerry A., and H. Allen Orr. 2004. *Speciation*. Sunderland, MA: Sinauer Associates.

De Queiroz, Kevin. 1999. "The General Lineage Concept of Species and the Defining Properties of the Species Category." In *Species: New Interdisciplinary Essays*, edited by Robert Wilson, 49–89. Cambridge: MIT Press.

Devitt, Michael. 2008. "Resurrecting Biological Essentialism." *Philosophy of Science* 75 (3): 344–82.

Dorr, Cian. 2010. Review of *Every Thing Must Go: Metaphysics Naturalized*, by James Ladyman and Donald Ross, with David Spurrett and John Collier. March 16 Revision. *Notre Dame Philosophical Reviews*. http://ndpr.nd.edu/news/24377-every-thing-must-go-metaphysics-naturalized/.

Dyke, Heather, and James Maclaurin. 2013. "What Shall We Do With Analytic Metaphysics? A Response to McLeod and Parsons." *Australasian Journal of Philosophy* 91 (1): 179–82.

Ereshefsky, Marc. 2007. "Foundational Issues Concerning Taxa and Taxon Names." *Systematic Biology* 56 (2): 295–301.

Ereshefsky, Marc, and Mohan Matthen. 2005. "Taxonomy, Polymorphism, and History: An Introduction to Population Structure Theory." *Philosophy of Science* 72 (1): 1–21.

Evans, Gareth. 1978. "Can There Be Vague Objects?" *Analysis* 38 (4): 208.

French, Steven, and Décio Krause. 2003. "Quantum Vagueness." *Erkenntnis* 59: 97–124.

Ghiselin, Michael. 1966. "On Psychologism in the Logic of Taxonomic Controversies." *Systematic Zoology* 26: 207–15.

Ghiselin, Michael. 1974. "A Radical Solution to the Species Problem." *Systematic Zoology* 23: 536–44.

Ghiselin, Michael. 1987. "Species Concepts, Individuality, and Objectivity." *Biology and Philosophy* 2 (2): 127–43.

Godfrey-Smith, Peter. 2006. "Theories and Models in Metaphysics." *The Harvard Review of Philosophy* 14: 4–19.

Griffiths, Paul E. 1999. "Squaring the Circle: Natural Kinds with Historical Essences." In *Species: New Interdisciplinary Essays*, edited by Robert A. Wilson, 209–28. Cambridge: MIT Press.

Haber, Matthew H. 2013. "Colonies Are Individuals: Revisiting the Superorganism Revival." In *From Groups to Individuals: Perspectives on Biological Associations and Emerging Individuality*, edited by Frédéric Bouchard and Philippe Huneman, 195–218. Cambridge: MIT Press.

Haber, Matthew H. 2015. "The Biological and the Mereological: Metaphysical Implications of the Individuality Thesis." In *Individuals Across the Sciences*, edited by Alexandre Guay and Thomas Pradeu, 295–316. New York: Oxford University Press.

Häggqvist, Sören. 2005. "Kinds, Projectibility and Explanation." *Croatian Journal of Philosophy* 5 (13): 71–87.

Hawley, Katherine. 1998. "Indeterminism and Indeterminacy." *Analysis* 58 (2): 101–6.

Heck, Richard. 1998. "That There Might Be Vague Objects." *The Monist* 81 (2): 274–96.

Hull, David L. 1977. "A Logical Empiricist Looks at Biology." *British Journal for the Philosophy of Science* 28 (2): 181–89.

Hull, David L. 1978. "A Matter of Individuality." *Philosophy of Science* 45: 335–60.

Hull, David L. 1989. *The Metaphysics of Evolution*. Albany: State University of New York Press.

Hull, David L. 1999. "On the Plurality of Species: Questioning the Party Line." In *Species: New Interdisciplinary Essays*, edited by Robert A. Wilson, 23–48. Cambridge: MIT Press.

Humphreys, Paul. 2013. "Scientific Ontology and Speculative Ontology." In *Scientific Metaphysics*, edited by Don Ross, James Ladyman, and Harold Kincaid, 51–78. Oxford: Oxford University Press.

Johnston, Mark. 1993. "Objectivity Refigured: Pragmatism Without Verificationism." In *Reality, Representation, and Projection*, edited by John Haldane and Crispin Wright, 85–130. New York: Oxford University Press.

Kitcher, Philip. 1989. "Some Puzzles About Species." In *What the Philosophy of Biology Is*, edited by Michael Ruse, 183–208. Dordrecht: Kluwer.

Kornblith, Hilary. 1993. *Inductive Inference and its Natural Ground*. Cambridge: MIT Press.

Koslicki, Kathrin. 2003. "The Crooked Path from Vagueness to Four-Dimensionalism." *Philosophical Studies* 114 (1–2): 107–34.

Kripke, Saul. 1980. *Naming and Necessity*. Cambridge, MA: Harvard University Press.

Ladyman, James, and Donald Ross, with David Spurett and John Collier. 2007. *Every Thing Must Go: Metaphysics Naturalized*. Oxford: Oxford University Press.

LaPorte, Joseph. 2004. *Natural Kinds and Conceptual Change*. Cambridge: Cambridge University Press.

Lewens, Tim. 2012a. "Pheneticism Reconsidered." *Biology and Philosophy* 27: 159–77.

Lewens, Tim. 2012b. "Species, Essence and Explanation." *Studies in the History and Philosophy of Biology and Biomedical Sciences* 43: 751–57.

Lewis, David. 1986. *On the Plurality of Worlds*. Oxford: Basil Blackwell.

Lewis, David. 1988. "Vague Identity: Evans Misunderstood." *Analysis* 48: 128–30.

Liebesman, David, and Matti Eklund. 2007. "Sider on Existence." *Noûs* 41 (3): 519–28.

Lowe, E. J. 1994. "Vague Identity and Quantum Indeterminacy." *Analysis* 54 (2): 110–14.

Maclaurin, James, and Heather Dyke. 2012. "What is Analytic Metaphysics For?" *Australasian Journal of Philosophy* 90 (2): 291–306.

Maddy, Penelope. 2007. *Second Philosophy*. New York: Oxford University Press.

Mallet, James. 1995. "A Species Definition for the Modern Synthesis." *Trends in Ecology and Evolution* 10 (7): 294–99.

Mayr, Ernst. 1959. "Darwin and the Evolutionary Theory in Biology." In *Evolution and Anthropology: A Centennial Appraisal*, edited by B. J. Meggars, 1–10. Washington, DC: Anthropological Society of Washington.

Mayr, Ernst. 1963. *Animal Species and Evolution*. Cambridge: Harvard University Press.

McLeod, Mike, and Josh Parsons. 2013. "Maclaurin and Dyke on Analytic Metaphysics." *Australasian Journal of Philosophy* 91 (1): 173–78.

Melnyk, Andrew. 2013. "Can Metaphysics Be Naturalized? And If So, How?" In *Scientific Metaphysics*, edited by Don Ross, James Ladyman, and Harold Kincaid, 79–95. Oxford: Oxford University Press.

Moore, Joseph G. 2008. "A Modal Argument against Vague Objects." *Philosophers' Imprint* 8 (12): 1–17.

Morganti, Matteo. 2013. *Combining Science and Metaphysics*. London: Palgrave–Macmillan.

Noonan, Harold W. 1995. "E. J. Lowe on Vague Identity and Quantum Indeterminacy." *Analysis* 55 (1): 14–19.

Paul, L. A. 2012. "Metaphysics as Modeling: The Handmaiden's Tale." *Philosophical Studies* 160: 1–29.

Putnam, Hilary. 1975a. "The Meaning of 'Meaning,'" in his *Mind, Language and Reality: Philosophical Papers*. Vol. 2, 215–71. Cambridge: Cambridge University Press.

Putnam, Hilary. 1975b. "Is Semantics Possible?," in his *Mind, Language and Reality: Philosophical Papers*. Vol. 2, 139–52. Cambridge: Cambridge University Press.

Quine, W. V. O. 1953. "On What There Is." Reprinted in his *From a Logical Point of View*, 1–19. Cambridge: Harvard University Press.

Quine, W. V. O. 1969. "Natural Kinds." Reprinted in his *Ontological Relativity and Other Essays*, 114–38. London: Columbia University Press.

Quine, W. V. O. 1981. "What Price Bivalence." *The Journal of Philosophy* 78 (2): 90–95.

Reydon, Thomas A. C. 2009. "How to Fix Kind Membership: A Problem for HPC Theory and a Solution." *Philosophy of Science* 76 (5): 724–36.

Richards, Richard A. 2010. *The Species Problem*. New York: Cambridge University Press.

Rieppel, Olivier. 2007. "Species: Kinds of Individuals or Individuals of a Kind." *Cladistics* 23: 373–84.

Ritchie, Jack. 2008. *Understanding Naturalism*. Durham: Acumen Publishing.

Ruse, Michael. 1987. "Biological Species: Natural Kinds, Individuals, or What?" *British Journal for the Philosophy of Science* 38: 225–42.

Sider, Theodore. 2001. *Four-Dimensionalism*. Oxford: Oxford University Press.

Sider, Theodore. 2003. "Against Vague Existence." *Philosophical Studies* 114: 135–46.

Sider, Theodore. 2009. "Against Vague and Unnatural Existence: Reply to Liebesman and Eklund." *Noûs* 43 (3): 557–67.

Slater, Matthew H. 2013. *Are Species Real?* London: Palgrave–Macmillan.

Slater, Matthew H. 2015. "Natural Kindness." *The British Journal for the Philosophy of Science* 66 (2): 375–411.

Slater, Matthew H. n.d. "Pluto and the Platypus: Tale of an Odd Ball and an Odd Duck." Unpublished manuscript.

Sober, Elliott. 1980. "Evolution, Population Thinking, and Essentialism." *Philosophy of Science* 47: 350–83.

Sokal, R., and P. Sneath. 1961. *Principles of Numerical Taxonomy*. San Francisco: Freeman.

Stanford, P. Kyle. 2006. *Exceeding Our Grasp: Science, History, and the Problem of Unconceived Alternatives*. New York: Oxford University Press.

Sterelny, Kim, and James Maclaurin. 2008. *What Is Biodiversity?* Chicago: University of Chicago Press.

Van Inwagen, Peter. 1988. "How to Reason About Vague Objects." *Philosophical Topics* 16 (1): 255–84.

Van Valen, Leigh. 1976. "Ecological Species, Multispecies, and Oaks." *Taxon* 25: 233–39.

Wheeler, Quentin D., and Rudolf Meier. 2000. *Species Concepts and Phylogenetic Theory*. New York: Columbia University Press.

Wilkerson, T. E. 1995. *Natural Kinds*. Brookfield: Ashgate Publishing Company.

Wilkins, John S. 2009. *Defining Species: A Sourcebook from Antiquity to Today*. New York: Peter Lang.

Williams, Neil E. 2011. "Putnam's Traditional Neo-Essentialism." *Philosophical Quarterly* 61 (242): 151–70.

Wilson, Robert A., Matthew J. Barker, and Ingo Brigandt. 2007. "When Traditional Essentialism Fails: Biological Natural Kinds." *Philosophical Topics* 35 (1/2): 189–215.

Winsor, Mary P. 2003. "Non-Essentialist Methods in pre-Darwinian Taxonomy." *Biology and Philosophy* 18: 387–400.

Winsor, Mary P. 2006. "Creation of the Essentialism Story: An Exercises in Metahistory." *History and Philosophy of the Life Sciences* 28 (2): 149–74.

4

No General Structure

C. Kenneth Waters

SCIENTIFIC METAPHYSICS CAN be described as an attempt to theorize about the fundamental nature of the world by drawing upon the best scientific theories of our day.[1] The fact that the best scientific theories of the past turned out to be wrong about the fundamental nature of the world poses a challenge. Why, skeptics ask, should we expect today's theories to provide a secure basis to inform us about the world's fundamentals? In this chapter, I propose that instead of drawing metaphysical conclusions by interpreting the most basic theories of science, we draw conclusions by analyzing how multifaceted practices of science work. This opens metaphysical inquiry to areas of science that are not organized around efforts to

[1] Niall Roe (2015) distinguishes between scientific metaphysics and the metaphysics of science. *Scientific metaphysics*, the topic of this chapter, can be thought of as metaphysically inquiry that is informed by science. One might pursue this approach and yet leave open the possibility that this is not the exclusive legitimate approach for metaphysics. The *metaphysics of science* can be thought of as inquiry into metaphysical assumptions, concepts, and principles embedded within science. This inquiry might be pursued using exclusively analytic approaches. For example, many metaphysicians assume that scientific reasoning critically relies on reasoning about dispositions. Much research on the metaphysics of dispositions, however, completely ignores contemporary science. Such research could be thought of as falling under the "metaphysics of science" even though it is not scientifically informed. Inquiry into metaphysical concepts embedded within science might also (or alternatively) be pursued by appealing to the findings or methodologies of contemporary science. See Stanford, this volume, chapter 6, for a somewhat different way of distinguishing between the metaphysics of science and scientific metaphysics. I prefer Roe's classification because it acknowledges that some metaphysicians are trying to investigate the metaphysical assumptions, concepts, and principles embedded within science from an analytic perspective.

articulate, develop, or extend the explanatory scope of a core theory.[2] I motivate and illustrate this approach for scientific metaphysics by analyzing what is one of the most successful scientific practices of today: the practice of using genetics to investigate, manipulate, and explain the workings of organisms.

This chapter begins by briefly discussing a traditional conception of scientific metaphysics. Next, it considers philosophical work aimed at identifying the fundamental nature of being in the biological world, such as being an organism, being a gene, or being a process of natural selection or genetic drift. Such work is not generally recognized by metaphysicians as an integral part of scientific metaphysics.[3] Perhaps this work does not seem sufficiently general because it is cast at the wrong scale. But I will argue that if we reconceive generality in terms of what one finds across different scales, then questions about what is true at intermediate scales (such as the scales of macromolecules, cells, organisms, and ecosystems) are not necessarily less general than questions about what is true at the largest or smallest scales. I call this the *generality across scales thesis*.[4] This sets the stage for developing the central argument of this chapter.

My central argument starts by applying the traditional approach of scientific metaphysics to the science of genetics. But instead of applying this approach to today's genetics, I first apply it to the genetics of the 1930s. I draw upon the best scientific theories of the era to theorize about the fundamentals of heredity, development, and evolution. In doing so, I consider the gene concept of classical genetics. Although genes were central to classical genetics, genes were entities with unknown structure and unknown proximate effect. Neither the what nor the how of their individual contributions were understood. Nevertheless, geneticists could accurately identify causal connections between genotypes and phenotypes in terms of *the difference principle*. This principle held in experimental contexts because geneticists deliberately simplified causal situations by standardizing environmental conditions and removing genetic differences that might affect the phenotypes under study. This manipulation and control is what made the practice of classical genetics possible, and I argue that it has important implications about how we should pursue scientific metaphysics.

---

2 This approach also opens the door to drawing metaphysical conclusions from areas of science whose practices are not organized by core theories that allegedly provide a comprehensive basis for explaining all phenomena within the practice's domain.

3 Many, or least some, philosophers of biology conducting this work do believe it is an integral part of metaphysics. My point is that outside philosophy of biology this work is not generally recognized as being integral to metaphysics.

4 The generality across scales thesis can be viewed as a challenge to Ladyman's idea that physics is "the science that holds at the widest range scales" (Ladyman, this volume, chap. 7).

The next step in my central argument is to show that although one could have been (and still can be) a realist about the basic theory of classical genetics, attempts to interpret this theory for the purpose of drawing conclusions about the fundamentals of heredity, development, and evolution were doomed to fail. Furthermore, understanding how the basic theory was deployed in the context of experimental practice reveals the futility of trying to draw conclusions about fundamentals from this best theory. But what about the best theory of today's genetics? Might the best theories of contemporary biology offer a more secure basis for identifying the fundamentals of heredity and development?

I examine a central concept of contemporary genetics (the molecular gene concept) and explain how it is employed in practice. Analyzing this concept in isolation of broader practice might lead one to conclude that it identifies the fundamental unit of heredity and development. Examining how the concept is employed in practice, however, reveals that it is not a category of being that "cuts nature at its joints." The problem with using this metaphor is that DNA has too many joints. I will argue that the practice of genetics utilizes a flexible gene concept because there is no overall structure, functional or causal, of the elements of DNA.

I expand my conclusion about genetics to motivate the *no general structure thesis*: the world lacks a general, overall structure that spans scales. It does not have a "*the* causal structure" that Salmon (1984) claims all explanations should fit into, and it lacks the kind of overall "definite and mind independent structure" that Psillos (1999) claims is an essential tenet of realism. This thesis is similar to the metaphysical views advanced by Cartwright (1983, 1999) and Dupré (1995), though I motivate it differently. I close the chapter by arguing that the *no general structure thesis* counts as significant metaphysics.

1. THE TRADITIONAL APPROACH OF SCIENTIFIC METAPHYSICS

Scientific metaphysics refers to the idea that metaphysics should be informed by science (Kincaid 2014). This characterization raises two difficult questions: (1) What is metaphysics? and (2) How should metaphysics be scientifically informed? One indication of the difficulty of answering the first question is the emergence of a subfield, meta-metaphysics, which is aimed at addressing questions about the nature of metaphysics as a subject area, including the question, "What is metaphysics?" (Chalmers, Manley, and Wasserman 2009; Tahko 2016). For the purposes of this chapter, I will assume that metaphysics is an area of philosophy that seeks answers to questions about the general nature of reality, the general nature of the things, kinds, processes, and complexities in the world. One of the questions raised by this chapter concerns how we should conceptualize the sense of "general" when we say that metaphysics seeks to answer questions about the general nature of reality.

The question of how metaphysics should be scientifically informed is also important, and this issue will be addressed in the present chapter. The literature on scientific metaphysics provides a somewhat ambiguous answer to this question. Some writings stress applying the *results* of science to answer metaphysical questions; other writings stress employing the *methods* of science. Philosophers emphasizing the results of science have focused on the most basic theoretical results of physics (e.g., Ladyman et al. 2007; Maudlin 2007; Callender 2011; and French 2014). This approach can be motivated by the idea that metaphysics is about the most fundamental features of the world, which is also what the theories of fundamental physics are about. The rough idea is that metaphysical conclusions should be drawn by interpreting the most basic theories of physics in order to determine what these theories say about the fundamentals of reality.

Theoretical results of biological sciences have not played the kind of key role that theoretical results of physics have played in work explicitly recognized as scientific metaphysics. (I explain the qualification "work explicitly recognized as" in section 2). Scientific metaphysicians who attend to the "special sciences" often assume that physics holds the trump cards for metaphysical questions and do not see a need to closely engage the actual theoretical contents of biological sciences to establish their basic metaphysical claims. For example, Ladyman et al. (2007) and Ladyman (this volume, chap. 7) argue that the core project of metaphysics is to unify the theoretical findings of special sciences with the fundamental theories of physics. Their casual engagement with the contents of biological theories is markedly different than their close and serious engagement with theories of physics, and biology plays a different kind of role than physics in their arguments. It is as if physics comes first, and biology had better agree. French (2014) gives more serious attention to the actual contents of biology, but its core thesis (ontological structural realism) is established on the basis of analyzing physics (beginning in chap. 1). Biology is not addressed until the last chapter of the monograph (chap. 12).

Writings about scientific metaphysics that emphasize the methods of science (rather than the results of science) often focus attention at the meta-level. For example, Maddy (2007) addresses questions about what methods metaphysics should employ, rather than directly employing those methods in metaphysical inquiry.[5] In contrast, the new experimental school of philosophy seeks to use scientific methods,

[5] The distinction breaks down if one conceives of scientific inquiry as an integral part of metaphysics, because under this conception the methods of science are (of course) being directly employed by scientific participants in the project of metaphysics.

usually from social psychology or cognitive science, to generate results that will directly bear on problems that are widely considered to be part of contemporary metaphysics. Philosophers have, for example, employed experimental methods to address problems of personal identity (e.g., Bruno and Nichols 2010), time (e.g., Paul 2010), causation (e.g., Hitchcock 2012), and ontology (e.g., Smith 1999). The distinction between scientific metaphysics that emphasizes applying the results of science and scientific metaphysics that emphasizes applying the methods of science is not a sharp one. Some scientific metaphysicians attempt to both engage the results of science and apply the methods of science. While I acknowledge the scientific methods approach, for the remainder of this chapter, I use the term *traditional scientific metaphysics* to refer to the variety of metaphysics that attempts to inform metaphysical inquiry by interpreting theoretical results of empirical investigations that have already been conducted by scientists.

Traditional scientific metaphysics, as specified above, typically involves interpreting the basic theories of physics to determine what these theories say about the fundamental reality of the world. But, as several authors have warned, this can be problematic since the basic theories of physics analyzed by metaphysicians are themselves grounded in metaphysical presuppositions (Sklar 1981; Hawley 2006; Chakravartty 2014). This raises the possibility that what is indirectly read off the fundamental theories of physics are the metaphysical biases of physicists, not empirically grounded metaphysical truths. Psillos (1999) responds that the key to avoiding this problem is to consider only those parts of scientific theories that are critical to the predictive success of the science. But I prefer another response. We should not conceive of scientific metaphysics as a purely empirical inquiry or as an inquiry that is informed by a purely empirical source. It is inquiry that, like science, involves an interplay or entanglement of non-empirical and empirical reasoning. The basic strategy underlying scientific metaphysics, I suggest, is based on the idea that the world provides constraints on scientific inquiry, and that philosophers can inform metaphysics by investigating these constraints.

2. QUESTIONS ABOUT BIOLOGICAL FUNDAMENTALS

Scientific metaphysics has directed attention primarily to questions about the "fundamental" features of the world, those features that are common to everything, always, everywhere. Work explicitly recognized as scientific metaphysics centers on fundamental physics because scientific metaphysicians assume that whatever is true of fundamental physics must be true of everything, always, and everywhere (and whatever is true of biology is not true of everything, always, and everywhere). However, there is considerable work in philosophy of biology that can be interpreted as a metaphysical quest to identify fundamentals. The fundamentals sought in these inquiries are the fundamentals of the living world (rather than of the universe at large). This work can be interpreted as scientific metaphysics even though it is not always recognized as such. Consider, for example, the questions:

What is life?
What is an organism (or a biological individual, or a Darwinian individual)?
What is a species?
What is natural selection? What is drift?
What is fitness?
What is group selection?
What is a population?
What is a gene?
What is genetic/genomic/biological information?
What is a biological signal?
What is a mechanism?
What is a function?

Philosophers of biology often analyze and critique scientists' answers to these questions with an eye toward identifying the fundamental reality of life, or the fundamental reality of being an organism, being a gene, or being a process of natural selection or drift. My claim is that many philosophers of biology (but not all) pursue these questions as if they are seeking fundamental answers to questions about ontological categories of being. One indication of this motivation is the frequent appearance of the term "fundamental" in their writings.

But, as mentioned in the introduction, this philosophical work in biology and philosophy of biology is not generally recognized as an integral part of scientific metaphysics (French [2014] is an exception). There are, perhaps, two reasons for this. One is that answers to these questions might not seem sufficiently general. For one thing, not everything in the universe is living or is even part of something that is living. So there is a scope issue. In addition, biology is cast at particular scales, so there is a scale issue as well. But if one reconceives generality in terms of what one finds at many different scales (some readers might say different "levels" [6]), then features found at intermediate scales (e.g., scales of macromolecules, cells, organisms,

[6] I put the term 'level' in quotation marks throughout this chapter because this terminology implies that nature is organized into levels. I do not assume that nature has an overall structure that is organized into levels, so I use the term 'scale.'

and ecosystems) are not necessarily less general than features found at the smallest or largest scales. The theory of quantum mechanics might describe general, structural features at very small scales, but it does not itself describe structural features that exist at larger, intermediate scales. For example, it does not describe structural features of the complexities in ecological dynamics or gene regulation that are being investigated by biologists. Perhaps the kinds of complexities that biologists deal with are quite general (even in non-living parts of the world) and representative of complexities throughout much of nature and across different scales. If so, then *knowledge about the form of complexities in biology would have a kind of generality that the basic theories of fundamental physics do not.* I call this admittedly contentious idea the *generality across scales thesis* and return to it in sections 7 and 8.

Perhaps another reason that metaphysicians do not generally recognize the questions presented above as metaphysical questions is that these questions do not yield neat, univocal answers. What is a species? According to many leading philosophers of biology, species are not any one kind of category or thing. As Ereshefsky (1998, 103) puts it, the "various 'taxa' called species lack a common unifying feature." And things do not parse more neatly at the level of genetics. Philip Kitcher is even more skeptical about genes than Marc Ereshefsky is about species: "a gene is whatever a competent biologist chooses to call a gene" (Kitcher 1992, 131).[7] Nevertheless, these questions are often posed by philosophers of biology in fundamental terms, as if what it is to be a species or a gene should come down to a few principled essentials. When it turns out that the diversity of life cannot be neatly partitioned into species or organisms, or when it turns out DNA cannot be neatly partitioned into genes, philosophers become skeptical about the reality of such kinds and search for other kinds that will hold up to their philosophical (i.e., fundamentalist) ideals. Hence the shift in philosophical attention from species to populations, from organisms to Darwinian individuals, and from genes to discrete functional units in DNA. It is as if philosophers of biology are framing questions in ways that call out for metaphysical answers; that is, answers that would provide a basis for drawing conclusions about the fundamentals of evolution, development, and life in general.

---

[7] Beatty (1995) convincingly argues that there is a reason why there are no fundamental laws of biology: biological generalizations are contingent on the historical process of evolution. Beatty's argument could be expanded to challenge the idea that there are fundamental biological entities, fundamental biological kinds, or fundamental biological processes as well. Whatever the explanation, unlike the basic theories in the parts of physics investigated by scientific metaphysicians, the basic theories across the biological sciences do not readily yield to the kind of fundamentalist interpretation associated with metaphysics. Of course, whether current or future theories will eventually yield to such an analysis is a contentious issue in philosophy of biology (e.g., Waters 2011; Okasha 2011).

It could be argued, by appealing to a reconceptualization of generality, that much research in philosophy of biology can be understood as scientific metaphysics. Under this interpretation, a central aim of philosophy of biology is to formulate basic questions in ways that answering them will identify the fundamentals of the biological world, for example the fundamentals of what it is to be a unit or process of heredity, development, or evolution.

3. WHAT IS A GENE? THE CLASSICAL GENE CONCEPT IN SCIENTIFIC PRACTICE

What is a gene? Contemporary geneticists employ multiple gene concepts that seem to offer conflicting answers. At one extreme is what I call the *classical gene concept*, which provides biologists with a blunt conceptual tool that works well in investigative and explanatory contexts in which precision is not available or useful (Waters 1994 and 2007). At the other extreme is what I call the *molecular gene concept*, which provides biologists with a remarkably precise and yet flexible tool for contexts in which precision is important (Waters 2000). The classical gene concept, as the name suggests, comes from classical genetics, the practice of genetics developed by T. H. Morgan and his collaborators in the late 1910s and 1920s. This practice continued until it was retooled to become contemporary genetics (Waters 2008a). In this section, I describe conceptual practice in classical genetics. In the next section, I consider how a traditional scientific metaphysician, if working in the 1930s, might have answered the question, "What is a gene?"

It is important to understand that the classical concept of the gene was grounded in experimental practice, not in abstract theorizing. Experimental practice in classical genetics involved strategically constructing breeding regimens that produced distinctive inheritance patterns. These patterns were explained by a central theory, called the transmission theory. This theory was constructed to apply to artificial experimental situations, and the experimental situations were constructed to instantiate the theory (Waters 2008b). This may appear to involve a problematic circularity or explanatory triviality. But these practices were aimed, not just at explaining transmission patterns, but at investigating a wide variety of biological processes, including hereditary, physiological, developmental, and evolutionary processes (Kohler 1994). When successful, this experimentation enabled biologists to manipulate processes they did not understand and to thereby reveal new information about these processes (Waters 2004).

While it is important to understand geneticists' theoretical explanations in the broader context of their experimental practice, one cannot understand this practice without analyzing how the theoretical/explanatory reasoning worked within it. Geneticists' reasoning about genes invoked a conceptual division between the internal genetic makeup of an organism, called its *genotype*, and its outward character, called its *phenotype*. They explained patterns of *phenotypic* transmission produced in their experiments by following the transmission of *genotypic* differences from one generation to the next, and then attributing the presence of alternative phenotypic traits to the presence of alternative genotypes (i.e., to the presence of alternative forms of genes). These explanations were based on the transmission theory, which included the idea that genes are located in linear fashion on chromosomes, principles about the transmission of genes that were grounded in an understanding of chromosomal mechanics, and the principle that differences in genes cause differences in phenotypes.

Let us consider the concept of the gene standing behind these explanations. As units passed down from generation to generation, genes were conceived as stable entities, capable of being replicated, located at designated positions in chromosomes. By at least the 1920s, most geneticists thought of genes as having physical structure, and it was the physical make-up of genes that was presumed to provide their stability. Practically nothing, however, was known about the internal structure of the gene until quite late in the development of classical genetics. The idea that there is a linear ordering of genes in chromosomes was essential to complex explanations of gene transmission, but this ordering does not imply anything about the structure of the genes themselves. Morgan and others frequently stated that their theory made two assumptions about the internal structure of genes: (1) gene structure is relatively stable, and (2) the structure of each gene is replicated before chromosomal division (e.g., Morgan 1926, 27). Muller (1922) pointed out that classical explanations positing spontaneous mutation made an additional assumption: (3) mutations in the structure of a gene are also replicated.

Just as Darwin's *Origin of Species* contained a scarcity of information about the nature of species and their origins, Morgan's *Theory of the Gene* had little to say about the structure of genes or their individual contributions to phenotypes. Genes could be speculatively related to the developing form of an organism, but the connection between genotype and phenotype was spelled out concretely only in terms of what I have called *the difference principle*: gene differences cause phenotypic differences within the genetic and environmental contexts of particular populations (Waters 1994, 2007). Strevens (this volume, chap. 2) describes a similarly simple model of predator/prey population cycles in ecology that also involves difference making. But he analyzes the ecological model in terms of an abstract explanatory principle ("enion probability analysis") that is ultimately grounded in a metaphysical theory, whereas I analyze models of transmission in terms of a concrete principle grounded in investigative practice. In the case of genetics, at least, it is important to keep in mind that the explanatory principle is easily and very accurately applied in experimental contexts because geneticists deliberately simplified experimental situations by standardizing environmental conditions and removing genetic differences that might interfere with the expression or transmission of genes being used in an experiment.[8]

It is easy to exaggerate the knowledge (or claims) of classical geneticists by focusing on theoretical explanations in an abstract context (rather than in the context of experimental investigation) or by citing geneticists' speculations. For example, classical geneticists have sometimes been accused by later scientists, philosophers, and historians of believing in a simple one gene/one character conception, a version of preformationism.[9] This misinterpretation might arise because geneticists' practices involved creating situations in which just one gene is the actual difference maker of a phenotypic difference in a laboratory population (Waters 2007). From a distance, it might appear that geneticists thought their explanations of experimental phenomena represented inheritance generally. But practicing geneticists understood that they were creating simple situations (which was obvious to them because they had to deliberately set up these situations to make their experiments work).

The relation between genotype and phenotype is not "one gene/one character," and classical geneticists knew this. As they clearly stated, one gene can affect a variety of characters, and a single character can be influenced by a number of different genes (genes at different loci). Geneticists knew, for example, that eye color in *Drosophila* is affected by mutations at many different loci; by 1915, Morgan's group had already identified mutations at twenty-five separate loci that affected eye color (Morgan et al. 1915, 208). They knew that these mutations generally affect several different characteristics. The white eye allele (located on the X chromosome), for example, was associated not just with white eyes, but with a colorless sheath of the testes, sluggish behavior, and perhaps a shortened life span as well. Experimental practice included protocols for selecting just one such phenotypic difference for experimental purposes. A one gene/one character conception of the gene makes sense only from a decontextualized, abstract perspective.

An examination of Morgan's sophisticated reflections on genotype/phenotype relations reinforces this interpretation. In *Theory of the Gene* (1926), Morgan reported

---

[8] I worry that Strevens's simple model cannot be applied to actual population cycles in an accurate way and that his analysis is too far removed from actual investigative practices in ecology to be useful. Stanford's critique seems to apply (Stanford, this volume, chap. 6).

[9] Examples of such exaggeration are too numerous to cite, but I take Moss (2003) and Keller's (2000) accounts of classical genetics as representative of how historians and philosophers exaggerate the claims of classical genetics (claims of which they are very critical). Keller (pers. comm.) makes the compelling argument that geneticists' speculations reveal biases that had significant impact on their research.

that embryology reveals that every organ of the body is the "culmination of a long series of processes," and he presumed that genes act on the steps along the way (1926, 306). If each step in the development of an organ is affected by many genes, he reasoned, then there could not be any single gene for the organ. Likewise, if one gene affects steps in the development of multiple organs, then there could not be any single organ associated with a gene: hence, the many-many relations. Morgan elaborated,

> Suppose, for instance, to take perhaps an extreme case, all the genes are instrumental in producing each organ of the body. This may only mean that they all produce chemical substances essential for the normal course of development. If now one gene is changed so that it produces some substance different from that which it produced before, the end-result may be affected, and if the change affects one organ predominantly it may appear that one gene alone has produced this effect. In a strictly causal sense this is true, but the effect is produced only in conjunction with all the other genes. (1926, 306)

Classical geneticists could only speculate about the immediate impact of genes (here, Morgan speculated that they produce chemical substances). This passage and Morgan's discussion of developmental processes suggest that the immediate impact of a gene is separated from characteristics such as eye color by a series of developmental processes that are also influenced by a number of other genes. This means that it would be impossible to specify a gene's contribution to phenotype in terms of characteristics such as eye color.

An abstract analysis of the transmission theory might make it appear that classical geneticists had a lot to say about genes, heredity, and development. But an examination of their theoretical knowledge in the broader context of experimental practice reveals that they did not have a lot to say about these matters. The real value of their science involved what they could manipulate and investigate, not what they could explain (and not even what their theory could *in principle* explain [see Waters 2004]).

My aim in this section is to make two points. First, the structure of the world that geneticists were manipulating and investigating was not directly reflected in the structure of their concepts and theories. Second, one gets a better sense of the complexity of the reality that geneticists were engaging when one analyzes their investigative and manipulative practices in light of their concrete, local aims (instead of analyzing their theoretical explanations abstractly—as if all we need to consider is their aim to explain inheritance patterns or their alleged aim to explain all development in terms of genes). Scientific metaphysicians interested in complex reality should focus on scientific knowledge (including theoretical knowledge) in the context of scientific practices (broadly speaking), not in an abstract context in which theories can be viewed separately from material practices designed to advance investigative and manipulative goals.

4. APPLYING THE TRADITIONAL APPROACH OF SCIENTIFIC METAPHYSICS TO THE BEST BIOLOGICAL THEORIES OF 1930S

Morgan was a meticulous empiricist, and he tried to avoid letting speculation enter into his scientific writings. But scientific metaphysics is by its nature speculative, and it is fair to ask how hypothetical scientific metaphysicians in the 1930s might have answered the question, "What is a gene?" Could their answer to this question provide a basis for answering questions about the fundamentals of heredity, development, and evolution?

Bold scientific metaphysicians of the 1930s (and it is fair to assume that these hypothetical metaphysicians would have been as bold as today's real ones) might have claimed that since presumably every stable inheritance pattern can be attributed to gene differences, that genes must be the *fundamental constituents of heredity*. In fact, many biologists, until quite recently, have written in textbook expositions and glossaries that genes are the "fundamental units of inheritance." What does it mean to say the gene is the *fundamental* unit of heredity? Presumably it means that if you could identify every gene and every difference in every gene, and if you could trace the transmission of each gene and each gene difference from one generation to the next, then you would have a comprehensive basis for understanding everything about heredity.

Scientific metaphysicians of the 1930s could have gone further. Since differences in genes cause phenotypic differences by affecting the way organisms develop, genes might also be the *fundamental units of development*. What would this mean? It would mean that if you could identify every gene and the functional role of every gene, then you would have a comprehensive basis for understanding everything in development. A scientific metaphysician of the 1930s would not need to speculate about the role of genes, but they could draw on the success of the transmission theory to claim that genes must have functional roles of some kind, and that fulfilling these roles must provide the fundamental basis underlying the development of organisms.

Although this might appear excessively speculative, consider the point that many contemporary metaphysicians take to be a marker of metaphysical significance: unification (e.g., Ladyman et al. 2007; Ladyman this volume, chap. 7). Identifying the gene as the fundamental unit of both heredity and development would unify their understanding of these phenomena. With the establishment of classical genetics, the science of heredity (genetics) and the science of development (embryology) had become disunified. But by going to the fundamentals, a metaphysician could have exclaimed, "This metaphysical interpretation reveals an underlying unity of nature!" This would be viewed as a basis for confidence. But why stop with the processes of inheritance and development?

Scientific metaphysicians of the 1930s might have wondered whether genes could also be *the fundamental units of evolutionary change*. Evolution involves the inheritance of changes in the forms of organisms over many, many generations. This must involve the inheritance of changes in the way organisms develop these forms. So, the metaphysical interpretation might go, genes must be the fundamental units of evolutionary change. Again, we could ask, what would this mean? Presumably, it would mean that if you could identify every gene and every difference in every gene that has occurred in evolutionary history, if you could identify the first appearance of each of these genes and gene differences, and if you could trace and explain the changes in frequencies of these genes and gene differences, then you would have a comprehensive basis for understanding evolution.

Scientific metaphysicians in the 1930s could have appealed to the latest developments in evolutionary theorizing and drawn upon the abstract theoretical work of Fisher, Haldane, and Wright to substantiate this view. The unification of fundamental principles of evolution with the fundamental unit of heredity and development could have made this interpretation of the best scientific theories of their day extremely appealing.

Of course, I am not advocating the idea that genes really are the *fundamental units* of heredity, development, or evolution, or even the more modest idea that scientists and philosophers of the 1930s had good reason for believing this. After all, section 3 shows that analyzing the classical concept of the gene in the context of scientific practice (rather than abstractly) reveals that there was good reason *not* to view classical genes as fundamental units of inheritance. The aim of this section is to show that today's favored approach to scientific metaphysics is unreliable. We can see this because if philosophers of the 1930s used this approach to identify the fundamentals of life, and if they analyzed the transmission theory and explanations in an abstract fashion (i.e., not in terms of the role they played in the context of investigative practice), they could have erroneously concluded that genes are the fundamental units of heredity, development, and evolution.

5. WHAT IS A GENE? THE MOLECULAR GENE CONCEPT IN SCIENTIFIC PRACTICE

How should contemporary scientific metaphysicians answer the question, "What is a gene?" Most philosophers weighing in on this question have concluded that trying to answer it is hopeless. For example, consider the prevalent idea that genes are units in DNA that "code for" polypeptides. (Some background: DNA is a macromolecule consisting of two complementary linear sequences of nucleotides; RNA is a macromolecule consisting of one linear sequence of nucleotides; a polypeptide is a molecule made up of a linear sequence of amino acids.) According to this prevalent idea, genes are subsequences of nucleotides in a strand of DNA. The basic story behind this conception goes as follows. A linear sequence of nucleotides in DNA comprises a gene, and this sequence is "transcribed" into a corresponding sequence of nucleotides making up an individual RNA molecule during RNA synthesis. Subsequently, the linear sequence of nucleotides in this RNA molecule is "translated" into a linear sequence of amino acids making up a polypeptide molecule during polypeptide synthesis. But it turns out that the syntheses of RNA and polypeptide molecules are far more complicated than this story suggests. These complications render this simplistic conception of genes ambiguous, vague, and exception-ridden.

On this basis, Evelyn Fox Keller (2000) argues that the term "gene" has outlived its usefulness, and many philosophers of biology agree with Keller. Much of the philosophical literature on this topic implies that the fundamental units of genetics exist at smaller scales and are more varied than suggested above. Some philosophers have argued that the real "molecular-level" units of genetics are not genes, but what were once thought to be parts of genes: promoters, enhancers, exons, and introns. Some philosophers have also argued that genes exist, but only at the "higher level" of classical genetics. Other philosophers have proposed novel gene concepts that seem to depart significantly from conceptual practice. For example, one idea is that genes are processes rather than entities (Griffiths and Neumann-Held 1999). But for the most part, philosophers have decided that today's science tells us that there is no such thing as a gene at the "molecular level."

I have argued that an analysis of how contemporary geneticists reason when they use the term "gene" reveals that they use a multiplicity of concepts (Waters 1994). Sometimes it is useful to be vague, and in such contexts biologists invoke a blunt concept akin to the gene concept of classical genetics (described in section 3). In other contexts it is important to be precise. When precision is important, biologists employ what I have called the *molecular gene concept* (Waters 1994, 2000).[10]

The molecular gene concept has placeholders. When the placeholders are filled, the concept can be used to pick out precise segments of DNA. So this concept is

[10] Paul Teller comments that chemistry employs different concepts of the atom in a similar way. Some concepts are useful in certain contexts because they are blunt. Other concepts are useful in other contexts because they are precise. Although what I say here is directly about genetics and allied sciences, I believe it applies to sciences more generally.

precise. But it is also flexible because the placeholders of the concept can be filled out in a multiplicity of ways, and the different instantiations that result can be used to pick out different segments of nucleotides in DNA. Some instantiations pick continuous segments; others pick out discontinuous segments. Some instantiations pick out segments that overlap segments picked out by other instantiations. The overall situation is very messy. But the flexibility of this concept enables biologists to pick out different precise segments of DNA that are relevant to different explanatory, investigative, and manipulative interests.

The molecular gene concept can be specified as follows:

> A gene *g* for linear sequence *l* in product *p* synthesized in cellular context *c* is a potentially replicating nucleotide sequence, *n*, usually contained in DNA, that determines the linear sequence *l* in product *p* at some stage of DNA expression.

The referent of any gene, *g*, is a specific sequence of nucleotides. The exact sequence to which a *g* refers depends on how the placeholders *l*, *p*, and *c* are filled out.[11] As figure 1 illustrates, this provides biologists with the conceptual means to pick out precisely what DNA segments determine different linear sequences in different stages and contexts of DNA expression.

The molecular gene concept is a remarkable conceptual tool. It gives biologists the flexibility they need to pick out DNA segments that line up with different causal chains (or processes) within the incredible complexities of DNA expression and development. It does so by providing the basis for partitioning the DNA molecule in a multiplicity of different ways.

In answer to the question, "What is a gene?," a contemporary scientific metaphysician adopting the traditional approach might answer that a gene is any segment of nucleotides that satisfies the analysis presented in this section. After all (trusting this analysis), this is what the best scientific theorizing of today employs as its gene concept.

6. PROBLEMS WITH APPLYING THE TRADITIONAL APPROACH OF SCIENTIFIC METAPHYSICS TO GENETICS

One problem with applying the traditional approach of scientific metaphysics to genetics is that the results are unstable. It leads to results that might seem plausible

---

11 The relevant sense of 'determine' is analyzed in Waters 2007.

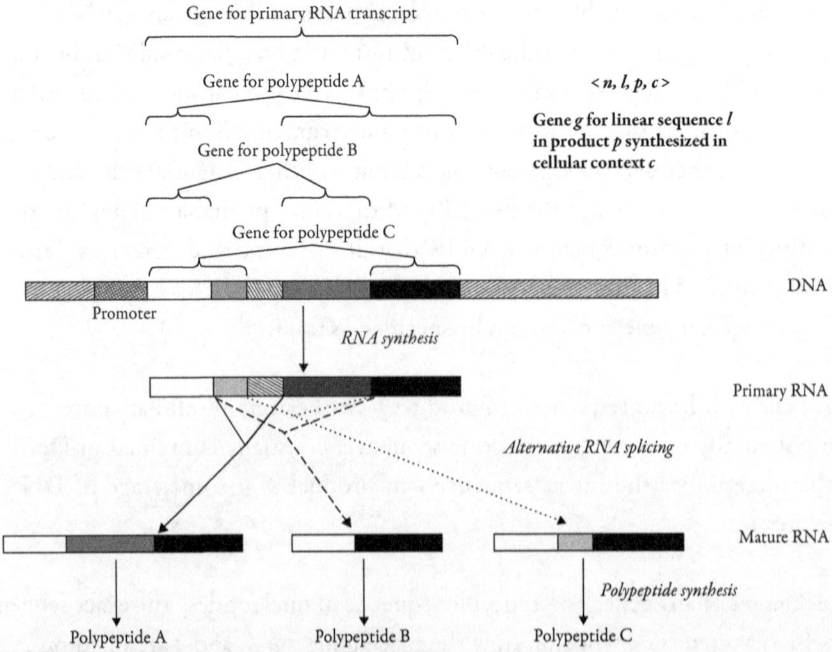

FIG. 1 The molecular gene concept enables biologists to partition DNA in multiple ways. With respect to this figure, if biologists are interested in the synthesis of the primary RNA transcript (see above), then the DNA segment covered by the bracket directly under 'gene for primary RNA segment' is the relevant gene (the continuous region of five segments in the DNA molecule starting with the solid white segment and ending with the solid black segment). If biologists are interested in the synthesis of polypeptide B occurring in some tissue at a particular stage of development, then the relevant gene is the discontinuous region marked by the brackets directly under 'Gene for polypeptide B' (the white and grey segments in DNA and not the segments between).

at one time, but turn out to be implausible afterward. For example, the fundamental interpretation of classical genetics set out in section 4 turns out to be simply false. Subsequent developments in genetics, developmental biology, and evolutionary biology have shown that genes are not *fundamental* units of heredity, development, and evolution. Genes are not the fundamental units of heredity and development because it is plainly false that knowledge about the functioning of each and every gene would provide the fundamental basis for explaining everything about heredity or development (Griffiths and Stotz 2013). Genes are not the fundamental units of evolution. Even if one could identify each and every gene that has occurred in evolutionary history, each and every difference in each gene that has occurred in evolutionary history, the first appearance of each of these genes and gene differences, and even if you could trace and explain the changes in frequencies of these gene differences, you would still not have a comprehensive basis for understanding all of evolution.

The attempt to interpret the best theories of the 1930s to identify the fundamentals of heredity, development, and evolution would have led to false conclusions (and the conclusions would not have been trivially false; they would have been false in significant ways). I wish to emphasize that the problem is with trying to interpret the transmission theory in a way that will reveal the *fundamentals*. The problem is not with the transmission theory as it was used in practice. The theoretical explanations, as sketched in section 3 (see Waters 2007 for more details), have withstood the test of time. Subsequent findings at the molecular level indicate that these explanations were correct. The transmissions of differences in functional units arranged linearly in chromosomes were indeed causally responsible for the patterns of phenotypic transmission produced in the laboratory. History does *not* undermine a realist interpretation of local explanations of classical genetics. But history does undermine an attempt to interpret the basic theory in a way that reveals the fundamental structure of the world (or the fundamental structure of inheritance, development, and evolution).

But perhaps the theories of genetics in the 1930s were not sufficiently developed to provide a basis for drawing stable metaphysical conclusions. Traditional scientific metaphysicians might respond that we should look to today's science, not yesterday's. And does not my analysis of the molecular gene concept support the idea that genes are fundamental units? After all, the analysis identifies a kind of internal structure and a kind of immediate functional role (in RNA synthesis) allegedly shared by all molecular genes. So this class of genes could be considered to be a natural (or fundamental) class. But this is a natural class (or kind) in a *weak sense*, in the sense of sharing an internal makeup and structure and having the same kind of proximate causal impact.

In a *stronger sense*, natural classes (or kinds) are supposed to provide the single correct parsing that reveals the joints of nature. Such classes would be *categories of being* that key into the fundamental structure of the world. If we looked at the gene concept out of the context of investigative and manipulative practices, it might appear to designate a natural class in this strong sense. But when we see how the concept is employed in practice, *and why it is a useful concept*, we see it designates a natural class only in the weak sense.

The molecular gene concept is useful not only because of its precision, but also because of its flexibility. The concept is relational. It is a *gene for* concept (Waters 2000). It can be applied differentially, to pick out different causal chains passing through DNA. These chains of causation extend from bewildering causal complexities of cellular functioning through different portions of DNA and then continue out into equally bewildering causal complexities of the cell. Multiple causal chains are entangled with one another and proceed through overlapping segments of DNA.[12] The molecular concept enables biologists to identify and manipulate individual chains in precise ways while other chains are stable (often systems are held stable via experimental manipulation). This enables biologists to slip and slide through causal entanglements that they do not understand in their natural complexity. In multicellular eukaryotes, differential splicing leads to different proteins in different functional contexts in different tissues at different stages of development. In some contexts, one set of exons plays a key role determining linear sequences; in other contexts other segments play key roles in determining linear sequences. The molecular gene concept enables biologists to specify the exons relevant to the synthesis of different proteins in particular cellular contexts without knowing much about the overall complexity of the natural situation.

The fact that this kind of conceptual practice works so well in genetics supports the idea that molecular genes are not a natural class in the stronger sense. That is, they do not provide a single correct parsing of DNA that reveals fundamental joints of nature. If molecular genes were a natural class in this strong sense, they would provide the unique division of DNA that would line up with the uniquely correct and comprehensive partition of development processes. Parsing DNA at its joints would line up with a parsing developmental at its joints. But geneticists have not identified such a canonical parsing. So, the metaphor of dividing nature at its joints does not seem to fit. There are simply way too many "joints" in DNA and way too many useful ways to parse the molecular and developmental processes. I have already explained that some parsings of DNA key into particular processes in particular tissues at particular developmental stages, other parsings key into different developmental processes in other tissues and/or other developmental stages. But the same is true from the other direction. Biologists have not found a parsing of genetic control processes that keys into a canonical set of joints in DNA. In sum, what they have found is that just as DNA complexly impinges on a causal mess, a causal mess also complexly impinges on DNA. That is why conceptual practice takes the form it does. Biologists have designed the molecular gene concept so they can navigate through a mess.

My concern is not that individuating genes in DNA depends on employing a relational concept. Rather, my point is that the way this relational concept is employed in practice indicates that the overall situation is far too messy to apply the metaphor of cutting nature at its joints. The remarkable feat of biologists is that they

---

12 Griffiths and Stotz (2013) supports this account of the causal situation, though it does not argue for the kind of metaphysical view advanced in this chapter. In fact, its argumentation seems to assume a much different epistemology and metaphysics than the one being developed in this chapter and the corpus of work this chapter draws upon (Waters 1994 through Waters 2014).

have succeeded in developing conceptual and technical tools that enable them to maneuver within bewildering messy complexities without having an overall theory or understanding of these complexities. My metaphysical claim is that scientific practices in genetics and allied sciences take this form because they are adapted to a reality that has no overall structure. The reality has lots of structure, but no overall structure.

7. DRAWING GENERAL CONCLUSIONS FROM CONCEPTUAL PRACTICE IN GENETICS

Scientific metaphysicians might respond to my skepticism about the idea that the intermediate-scale world that biologists engage lacks an overall structure by retreating from biology. "Perhaps biology is not ready for metaphysics. Perhaps geneticists haven't found nature's joints. But the joints *must* exist. If not in DNA, then elsewhere." I disagree. Genetics is sufficiently advanced to draw important metaphysical conclusions. But to draw them we must broaden our attention from a narrow focus on its theories and theoretical concepts, to how its theories and concepts are used in the context of broader investigative and manipulative practices. In addition, we should not assume at the outset that the world must have an overall structure, or that living systems and processes must have an overall structure at the scale being manipulated and investigated by biologists.

The practice of genetics enables biologists to control and manipulate a wide variety of phenomena and to gain new knowledge about the entities and processes of life. If the metaphysical ideas set out in section 4 were true, then the practice of contemporary genetics would presumably take a different form than it does. It would be based on a gene concept that provided a canonical parsing of DNA. But contemporary gene concepts do not provide such a parsing (or imply that such a parsing exists *even in principle*). Instead, biologists rely on gene concepts that enable them to parse DNA in a multiplicity of ways. Why does their practice take this form? The simple answer is "because it works." Practice has been adapted to work in the reality of the world that biologists are engaging.

A metaphor might be helpful here.[13] Biologists using genetics to investigate the workings of organisms are like newcomers trying to navigate in a strange city. Suppose they enter the city of Arles in southern France. The streets of Arles lack a general, overall structure. You can learn to navigate in one small part of the city, but that does not help you anticipate the layout of another part of the city. Contrast

[13] I thank John Norton for suggesting this metaphor.

this with newcomers arriving in the western Canadian city of Calgary. The streets of Calgary are laid out in a grid with four quadrants. "Streets" are oriented north/south (they run from north to south or from south to north), and "avenues" are oriented east/west. Streets are numbered sequentially in each quadrant, starting from the center of the city. For example, streets in the northwest quadrant are named 1 Street NE, 2 Street NE, and 3 Street NE. Avenues are named in the same fashion, for example 1 Avenue NE 2, Avenue NE, and 3 Avenue NE.

The best strategy for learning to navigate in cities like Calgary is to key into the overall structure. Exceptions to the grid structure exist near rivers and railroad tracks, but knowing there is some general, overall structure will facilitate newcomers' exploration of the city. Newcomers investigating Washington, DC would also be well-served by a strategy that rests on the assumption that there is some overall structure of streets, even though the structure of streets in DC is quite different than the structure of streets in Calgary. But newcomers to Arles should strategize their investigation differently because there is no overall structure to the street layout in Arles. My claim is that the domains being investigated by biologists using genetics are more like Arles than Calgary. We can infer this from the nature of the practices used in genetics. The world within organisms, like the street layout of Arles, is a mess. The practice of genetics has been adapted to navigate through the mess; the molecular concept of the gene is designed to enable biologists to investigate and control causal pathways within a messy, dynamic entanglement of interacting causal chains, an entanglement that they do not understand in its complexity.[14]

The idea that the reality within organisms lacks an overall structure can be generalized. It implies that the world in its entirety lacks a general overall structure. That is, it suggest that the world lacks a "the structure" that spans scales. But perhaps the domain (part and scale of the world) investigated by genetics is an exception. Perhaps the world is like the street layout of Calgary, and the living portion of the world at the scale being investigated by geneticists is akin to the portion of Calgary located near the Elbow River, where the grid breaks down and the street layout is messy.

In section 2, I described the generality across scales thesis, which states that features of the world at intermediate scales (such as the scales of macromolecules, cells, organisms, and ecosystems) are not necessarily less general than features at the smallest and largest scales. The generality across scales thesis implies that the feature of having no overall structure could obtain quite generally across scales being

[14] The distinction and metaphor presented here is somewhat similar to Feynman's distinction between Greek and Babylonian approaches to theorizing (1967) as discussed by Wimsatt (2007). But Wimsatt's rich and provocative discussion seems to assume that the world itself has a fundamental structure even though physical theories do not.

investigated by biological and social scientists, and by physical scientists in many fields as well. Why should we think that the universe has structure at every scale or that the principles (or laws) at the smallest scales provides an overall structure at larger scales?[15] Is the structure of the world a fractal?

There is much work to be done to clarify the widespread assumption that the world has a "*the* structure" and to articulate the contrary no general structure thesis. This work will require drawing upon traditional work in metaphysics as well as carefully analyzing a greater breadth of scientific practices across a range of sciences, biological, social, and physical. This chapter offers a start by arguing for the idea that the reality engaged by geneticists lacks an overall structure and by raising the question of whether the widespread metaphysical assumption that the world has a "the structure" is mistaken.

8. DOES THE NO GENERAL STRUCTURE THESIS COUNT AS SIGNIFICANT METAPHYSICS?

Does the claim that the world lacks a general, overall structure that spans scales count as a *significant* metaphysical idea? I close this chapter by responding to four objections to the supposition that it does. But before proceeding with the first objection, it is worth pointing out that a denial of a metaphysical thesis is itself a metaphysical thesis. The assumption that the world has a "the structure" across scales, which is widely held by philosophers of science, is a metaphysical thesis. Hence, claims that deny this assumption are also metaphysical. So the claim that the reality within organisms lacks an overall structure at the scale at which geneticists engage, and more generalized versions of this claim, must also be metaphysical.

> Objection 1: The no general structure thesis is too skeptical to count as metaphysics. After all, it amounts to an antirealist view about science, so it must be anti-metaphysics.

This chapter does not advance an antirealist view of science. Section 3 sketches classical geneticists' explanations of phenotypic transmission patterns for two purposes. The first purpose is to show that the theoretical explanations of classical genetics

---

[15] The significance of the idea that the science of fundamental physics "holds" across scales, often expressed by philosophers of physics (e.g., Ladyman, this volume, chap. 7), is unclear. Does it imply that the laws or principles of fundamental physics structure the world across scales? This depends upon how we interpret 'holds.' I believe there is a way to interpret the idea that a set of principles "holds across scales" that is consistent with the no generality across scales thesis. This is an idea worth exploring, but I do not have the space to do so here.

have stood the test of time. We should be realists about the geneticists' theoretical claims that gene differences caused phenotypic differences in the experimental contexts they constructed and that patterns of phenotypic transmission produced in experiments resulted from the transmission of gene differences from one generation to the next. Classical geneticists were right about the causal relationship between genotype and phenotype in their experiments, they were right about gene differences being located in linear fashion in chromosomes, and they were right about the roles that chromosomal mechanics in meiosis played in producing the patterns of gene transmission they produced in the laboratory.

The second purpose for sketching the theoretical explanations in classical genetics is to separate this realist account of the transmission theory from a fundamentalist interpretation of this theory (the former is presented in section 3, the latter in section 4). I agree with Cartwright (1999, 23) that fundamentalism, not realism, is the problem.

We should also be realists about the central theory of contemporary genetics, but again we should separate our realism from a fundamentalist interpretation of the theory. Scientific metaphysics should proceed from an analysis of the form that practice in genetics takes, not from an analysis of its core theoretical concepts removed from the context of that practice. Simply put, metaphysics should be practice centered, not theory focused.[16] But practice-centered metaphysics does not ignore theory, and it does not adopt an anti-realist attitude toward those theories. Hence, this metaphysics is not based on antirealism.

> Objection 2: Can we really draw a conclusion about metaphysics from scientific practice? Does not this analysis simply reveal epistemological or cognitive limitations?

This objection expresses a worry that applies to all metaphysics. When metaphysicians employ methods of analytic philosophy to investigate how the world must be, might they just be investigating how *we* must conceive the world because of our cognitive limitations? When traditional, scientific metaphysicians interpret the theories and explanations of physics to investigate "the structure" of the world, might they just be reinforcing scientists' biases or theories that are partly shaped by our limited cognitive abilities? Might the apparent mess of causal interactions within organisms being investigated using genetics have an overall structure that humans cannot recognize because of cognitive limitations? Yes, yes, and yes.

Metaphysics is risky business. Scientific metaphysics is based on the idea that appealing to science to inform metaphysics will decrease the risks. Appealing to a

[16] For more on the distinction between practice-centered and theory-focused, see Waters 2014.

broader analysis of science, one that examines the use of concepts and theories in the context of investigative, manipulative, and local explanatory practices will better inform metaphysics. I maintain that informing metaphysics by analyzing the concepts and theories of science from a narrow, theory-focused perspective is riskier than informing metaphysics by analyzing the investigative and manipulative practices in which these concepts and theories are employed.

> Objection 3: This conclusion is not sufficiently general to count as metaphysics.

A preliminary response to this objection is given in section 2. But it is worth restating with respect to the no general structure thesis. A critic advancing this objection might be willing to grant that the causal processes within an organism lack an overall structure at the scale at which geneticists engage. But the critic could still object, "So what if the world at this scale within organisms is a mess? This does not mean that the nonliving world at this scale is a mess (scope issue). And it does not even mean that the world within the organism is a mess (scale issue). To determine whether the world is a mess, one needs to consider everything in the whole world, not just some entities within it (organisms). And to consider the world at this universal scope, one must consider it at the appropriate scale, a much smaller scale than the scale of causal interaction being investigated by geneticists. Fundamental physics is the only discipline that can reveal whether the world has an overall structure and what that structure might be."

In section 2, I respond to this kind of objection by suggesting that we should adopt a different sense of general. Instead of thinking of general in terms of applying to everything at some one scale, we should conceive of generality across different scales. Claims about an overall structure existing at very small scales are not *necessarily* informative about whether there are overall structures at larger scales. The *generality across scales thesis* suggests that the no overall structure idea applies quite generally across scales being investigated by biological and social scientists, and by physical scientists in many fields as well. Metaphysics can (and should) be thought of as concerning what is true across many scales, *especially* across scales we directly engage and experience as human beings.

> Objection 4: This conclusion does not count as a significant metaphysical thesis because it is trivial.

Stanford (this volume, chap. 6) critiques traditional scientific metaphysics and argues that it adds nothing of value to philosophy of science. I believe Stanford's critique raises an important question: what is scientific metaphysics good for? Being informed by science, by itself, does not make a metaphysics important. Metaphysics should not only be informed by science; metaphysics should be informing science, the rest of philosophy, and society as well. I conclude this chapter by arguing that the no general structure thesis is a metaphysical doctrine of real importance.

With respect to philosophy, the no general structure thesis can inform our epistemology of science. Consider philosophy of biology. This thesis implies that the questions set out in section 2 should *not* be interpreted as fundamental questions. The distinction drawn with respect to classical genetics should be drawn with respect to these questions as well. In the case of classical genetics, the science provided accurate descriptions and explanations, but it should not be interpreted to yield a fundamental account of heredity, development, or evolution. Such interpretations rest on an assumption that the processes of heredity, development, and evolution have some fundamental structure such that it must be possible to find a basis of scientific understanding that keys into that structure. But the no general structure thesis implies that our interpretations of scientific knowledge should not be premised on the faith that such structures exist. This thesis has important implications for how philosophers of biology should interpret and answer questions such as "what is a species?," "what is an organism?," or "what is a Darwinian population?" In addition, the no general structure thesis has implications for how we should interpret and answer meta-scientific questions such as "what is a natural kind?"

The no general structure thesis can inform science as well. The quest for a comprehensive and unified, or even integrated, explanatory perspective can be a useful heuristic, but it should be viewed as a heuristic, not as *the* aim of science. Practicing geneticists were right not to follow philosophers when we obsessed about questions such as "what is fitness?" or "what is a gene?" It might be heuristically useful to ask such questions, but when it is not useful, such questions can simply be dropped and the project of investigating and manipulating nature can resume. A philosophy of biology that asked, "What ways of conceiving of biological individuals could be useful?" and "in what contexts and for what purposes would they be useful?" would be much more informative to biologists than simply asking, "What is a biological individual?." The no general structure thesis is important to science. A philosophy of science that adopted this thesis would be more useful to scientists.

The no general structure thesis is far from trivial. It has important implications for how scientists should conduct their investigations and how we (philosophers of science) should conduct ours. But should not metaphysics do more? Should it not also inform society about how to conduct its affairs? The no general structure thesis can deliver on this desideratum as well. It suggests that society should admire science for its secure knowledge about the world, but that society should not interpret that knowledge as if it depended upon or provided an understanding of fundamentals.

Research in genetics should be supported and funded, but not because it will "decode life" or reveal the magic keys to unlocking solutions to the complex problems societies face. The no general structure thesis applies generally across scales at which we experience the world. It has implications for how philosophers should understand scientific knowledge and for how scientists construct such knowledge. And it can inform the public understanding, support, and use of science. Such a thesis counts as significant metaphysics.

ACKNOWLEDGMENTS

Thanks to audiences at the Universities of Toronto, Vienna, Cologne, and California-Irvine where I presented talks on this topic. I have received useful feedback from too many individuals to name. But I would like to thank Will Bausman, Rick Creath, Andreas Hütteman, Marie Kaiser, Kyle Stanford, Marcel Weber, Bill Wimsatt, referees of Oxford University Press, and especially Alan Love for helping me along the way. Colleagues at the University of Calgary including Soohyun Ahn, Adrian Currie, Brian Hanley, Noa Latham, Niall Roe, and especially Marc Ereshefsky provided useful advice. Paul Teller offered valuable insights on the penultimate draft of this chapter. Finally, I would like to thank Matt Slater for his patience and encouragement. The research and writing of this chapter was supported in part by a grant from the John Templeton Foundation: #50191; *From Biological Practice to Scientific Metaphysics*; http://biological-practice-to-metaphysics.org.

REFERENCES

Beatty, John. 1995. "The Evolutionary Contingency Thesis." In *Concepts, Theories, and Rationality in the Biological Sciences*, edited by Gereon Wolters and James G. Lennox, 45–81. The Second Pittsburgh-Konstanz Colloquium in the Philosophy of Science. Pittsburgh: University of Pittsburgh Press.

Bruno, Michael, and Shaun Nichols. 2010. "Intuitions About Personal Identity: An Empirical Study." *Philosophical Psychology* 23 (3): 293–312.

Callender, Craig. 2011. "Metaphysics and Philosophy of Science." In *The Continuum Companion to Philosophy of Science*, edited by Stephen French and Juha Saatsi, 33–54. London: Continuum International Publishing Group.

Cartwright, Nancy. 1983. *How the Laws of Physics Lie*. New York: Oxford University Press.

Cartwright, Nancy. 1999. *The Dappled World: A Study of the Boundaries of Science*. Cambridge: Cambridge University Press.

Chakravartty, Anjan. 2014. "On the Prospects of Naturalized Metaphysics." In *Scientific Metaphysics*, edited by Don Ross, James Ladyman, and Harold Kincaid, 27–50. Oxford: Oxford University Press.

Chalmers, David, David Manley, and Ryan Wasserman, eds. 2009. *Metametaphysics: New Essays on the Foundations of Ontology*. Oxford: Oxford University Press.

Dupré, John. 1995. *The Disorder of Things: Metaphysical Foundations of the Disunity of Science*. Cambridge, MA: Harvard University Press.

Ereshefsky, Marc. 1998. "Species Pluralism and Anti-Realism." *Philosophy of Science* 65: 103–20.

Feynman, Richard. 1967. *The Character of Physical Law*. Cambridge, MA: MIT Press.

French, Steven. 2014. *The Structure of the World: Metaphysics and Representation*. Oxford: Oxford University Press.

Griffiths, Paul E., and Eva M. Neumann-Held. 1999. "The Many Faces of the Gene." *BioScience* 49 (8): 656–62.

Griffiths, Paul E., and Karola Stotz. 2013. *Genetics and Philosophy: An Introduction*. Cambridge Introductions to Philosophy and Biology. Cambridge: Cambridge University Press.

Hawley, Katherine. 2006. "Science as a Guide to Metaphysics?" *Synthese* 149: 451–70.

Hitchcock, Christopher. 2012. "Portable Causal Dependence: A Tale of Consilience." *Philosophy of Science* 79 (5): 942–51.

Keller, Evelyn Fox. 2000. *Century of the Gene*. Cambridge: Harvard University Press.

Kincaid, Harold. 2014. "Introduction: Pursuing a Naturalist Metaphysics." In *Scientific Metaphysics*, edited by Don Ross, James Ladyman, and Harold Kincaid, 1–26. Oxford: Oxford University Press.

Kitcher, Philip. 1992. "Gene: Current Usages." In *Keywords in Evolutionary Biology*, edited by Evelyn Fox Keller and Elisabeth A. Lloyd, 128–31. Cambridge, MA: Harvard University Press.

Kohler, Robert E. 1994. *Lords of the Fly: Drosophila Genetics and the Experimental Life*. Chicago: University of Chicago Press.

Ladyman, James, and Don Ross, with David Spurett, and John Collier. 2007. *Everything Must Go: Metaphysics Naturalized*. Oxford: Oxford University Press.

Maddy, Penelope. 2007. *Second Philosophy*. Oxford: Oxford University Press.

Maudlin, Tim. 2007. *The Metaphysics within Physics*. Oxford: Oxford University Press.

Morgan, Thomas Hunt. 1926. *The Theory of the Gene*. New Haven, CT: Yale University Press.

Morgan, Thomas Hunt, Alfred Henry Sturtevant, Hermann Joseph Muller, and Calvin Blackman Bridges. 1915. *The Mechanism of Mendelian Heredity*. New York: Henry Holt and Company.

Moss, Lenny. 2003. *What Genes Can't Do*. Cambridge, MA: MIT Press.

Muller, Hermann Joseph. 1922. "Variation Due to Change in the Individual Gene." *American Naturalist* 56: 32–50.

Okasha, Samir. 2011. "Reply to Sober and Waters." *Philosophy and Phenomenological Research* 82 (1): 241–48.

Paul, Laurie A. 2010. "A New Role for Experimental Work in Metaphysics." *Review of Philosophy and Psychology* 1 (3): 461–76.

Psillos, Stathis. 1999. *Scientific Realism. How Science Tracks Truth*. London: Routledge.

Roe, Niall. 2015. "Scientific Metaphysics or the Metaphysics of Science?" Seminar paper, University of Calgary, 2015.

Salmon, Wesley C. 1984. *Scientific Explanation and the Causal Structure of the World*. Princeton, NJ: Princeton University Press.

Sklar, Lawrence. 1981. "Time, Reality, and Relativity." In *Reduction, Time, and Reality*, edited by Richard Healey. Cambridge: Cambridge University Press.

Smith, Barry. 1999. "Ontology with Human Subjects Testing: An Empirical Investigation of Geographic Categories." *American Journal of Economics and Sociology* 58 (2): 245–72.

Tahko, Tuomas E. 2016. *An Introduction to Metametaphysics*. Cambridge: Cambridge University Press.

Waters, C. Kenneth. 1994. "Genes Made Molecular." *Philosophy of Science* 61: 163–85.

Waters, C. Kenneth. 2000. "Molecules Made Biological." *Revue Internationale de Philosophie* 4/2000 (214): 539–64.

Waters, C. Kenneth. 2004. "What Was Classical Genetics?" *Studies in History and Philosophy of Science* 35 (4): 783–809.

Waters, C. Kenneth. 2007. "Causes That Make a Difference." *The Journal of Philosophy* 104 (11): 551–79.

Waters, C. Kenneth. 2008a. "Beyond Theoretical Reduction and Layer-Cake Antireduction: How DNA Retooled Genetics and Transformed Biological Practice." In *Oxford Handbook to the Philosophy of Biology*, edited by Michael Ruse, 238–62. New York, NY: Oxford University Press.

Waters, C. Kenneth. 2008b. "How Practical Know-How Contextualizes Theoretical Knowledge: Exporting Causal Knowledge from Laboratory to Nature." *Philosophy of Science* 75 (5): 707–19.

Waters, C. Kenneth. 2011. "Okasha's Unintended Argument for Toolbox Theorizing." *Philosophy and Phenomenological Research* 82 (1): 232–40.

Waters, C. Kenneth. 2014. "Shifting Attention From Theory to Practice in Philosophy of Biology." In *New Directions in the Philosophy of Science*, edited by M. C. Galavotti, D. Dieks, W. J. Gonzalez, S. Hartmann, T. Uebel, and M. Weber, 121–39. Berlin: Springer International Publishing.

Wimsatt, William C. 2007. *Re-Engineering Philosophy for Limited Beings: Piecewise Approximations to Reality*. Chicago: Chicago University Press.

5

An Empiricist's Guide to Objective Modality

Jenann Ismael

1. MODALITY

Modality is a sticking point for empiricists. Some of them (e.g., Ladyman 2004) recognize that science comes with a heavy dose of modal commitment—a whole slew of beliefs not only about how things are, but how they might have been, could have been, or would have been had things been otherwise—and accept modality on the strength of their commitment to science. But others shun it on the grounds that talk of non-actual possibilities is epistemologically and metaphysically suspect.[1] If we take the semantics of modal belief at face value, moreover, it seems that we are committed to the existence of non-actual possible worlds, and it has never been clear what these are, or how we could know about them. So understanding modality is a matter of some urgency for philosophers of science. On the one hand, there is a near-universal recognition that modality is central to science. Science cares not only about the pattern of actual events but also what they reveal about the modal substructure behind the phenomena. On the other hand, there is among many a desire to reject metaphysical commitment to non-actual, possible worlds. To satisfy that desire while preserving scientific practice, one needs an account of modality that makes modality immanent in the actual world, i.e., one that does not take the semantics of modal belief at face value.

[1] Van Fraassen takes rejection of modality to be one of the defining features of empiricism: "To be an empiricist is to withhold belief in anything that goes beyond the actual, observable phenomena *and to recognize no objective modality in nature*" (1980, 202; emphasis mine).

Early attempts to remove metaphysical commitment took the form of attempts to reduce modal concepts to non-modal ones. Views of this form hold that modal concepts do not add anything to a description of the pattern of actual events. Such views run into difficulties, however getting the contents of modal belief correct. In this paper, I defend an empiricist account of modality that keeps a substantive account of modal commitment, but throws out the metaphysics. I suggest that if we pair a deflationary attitude toward representation with a substantive account of how scientific models are constructed and put to use, the result is an account that deflates the metaphysics of modal commitment without deflating the content of modal claims.

In section 2, I review the recalcitrant stumbling block for reduction. In section 3, I look at the function of models and the practical and epistemic role that modal structures play.[2] In section 4, I introduce a generic account of modal structures as partially prepared solutions to frequently encountered problems (PPS's to FEP's). In section 5, I compare my account of laws and chances with the account that comes out of David Lewis's Best Systems Analysis. In section 6, I suggest that the account avoids the pitfalls of both reification and reduction.

2. THE PERSISTENT STUMBLING BLOCK FOR REDUCTION

Modal notions enter science with concepts like cause, natural law, probability, dispositions, and capacities.[3] These notions all have a modal component. To say that L is a law is to say more than that things always happen in accord with L. It is to say that things happen in accord with L in every physically possible world (or that every physically possible world satisfies L). To say that a certain type of event (e.g., a coin flip coming up heads, or an atom decaying within a given time frame) has a good chance of occurring is not to say that it *does* happen, but to say that it happens in a large measure of physically possible worlds. Similar things can be said about causes, capacities, and dispositions. The truth conditions of claims about these structures, on standard construals, make reference to possible worlds.[4]

[2] I use "structure" in a generic way to refer to any property or relation. For examples of discussion in this mold, see Price (2011) and Kment (2014). I would not follow Kment in calling it an external standpoint, because it will itself *use* modal concepts. I explicitly deny the possibility of stepping outside the practice, as though we could describe or conceptualize or come to understand anything without using modal notions. The idea is rather to take a side-on view of the practice in which these concepts arise that is internal to a fully articulated scientific picture of the world. Doing this kind of genealogy for everyday modal belief is more difficult, since the epistemology of everyday belief is less explicit and systematic, and since the logic of common sense is less regimented than that of scientific discourse.

[3] I have benefitted greatly from audiences in Toronto and Rutgers, from discussion and correspondence with Barry Loewer, and from comments by Katherine Brading and Martin Jones that helped the paper immeasurably.

[4] Possible Worlds Semantics, which analyzes modal beliefs as beliefs about possible, non-actual worlds, has greatly clarified the formal and logical properties of modal thinking. According to Possible Worlds Semantics,

Here and throughout, I use laws and chances as exemplars of modal structures in science because they appear explicitly in fundamental theories, and because there are particularly well-developed discussions of their ontological status. I use the term "structure" generically to refer to any property or relation defined on our models of the world. According to this usage, dispositions, causes, and capacities provide other examples of modal structures.

Dynamical laws typically take the form of differential equations that give the rate of change for a quantity at a point. They entail the existence of regularities, but cannot be simply identified with regularities because the notion of law recognizes the possibility of regularities that are not laws. So, for example, we can easily write down a solution to the Newtonian equations of motion in which all of the birds in flight over Australia change direction as the clock strikes noon every day in Sydney. If this actually occurred, it might raise our suspicions that there were some laws that hadn't been identified, but it is possible that this is a merely accidental regularity, and the possibility that it is an accidental regularity is provided for by the Newtonian laws. It does not help to place restrictions on which *kinds* of regularities get counted as laws. Whatever condition one places on which regularities get counted as laws, it will always be possible to find solutions for a set of laws in which there are accidental regularities of the relevant kind.[5]

A very similar point can be made about chances. Chances are single-case probabilities represented by a real valued function over space-time.[6] The chance of a particular *a* that is *b* (e.g., a particular flip of a coin that lands heads) is derived from the indefinite probability of *a*/*b* (the indefinite probability that a flip of the same, or physically indistinguishable coin lands heads). And indefinite probabilities are connected to frequencies via a number of theorems of different strengths (the weak and strong laws of large numbers). So the link between chances and categorical facts goes by way of a link to indefinite probabilities, and indefinite probabilities are connected to frequencies in a way that suggests that they are not entirely distinct existences. The relationship between chances and frequencies is very like the relationship between laws and regularities. As with laws, views that try to reduce

---

to say that something might have happened is to say that it happens in some possible world. To say that something could not happen is to say that it happens in no possible world. To say that something would happen under conditions C is to say that it happens in the nearest possible world in which C obtain.

[5] This is a complex issue that has been rather heavily discussed in the literature. For an overview of the problems faced by reductive accounts of law, see Carroll (2016). For an overview of the problems faced by reductive accounts of chance, see Hajek (2011).

[6] In referring to chances as single-case probabilities, I am following the tradition of Lewis (1980). See also Bigelow, Collins, and Pargetter (1993) and Hall (1994). The philosophical discussion of chance, however, is fraught by ambiguous usage. Chances are sometimes identified with the indefinite probabilities described above.

probabilities to frequencies face a number of well-known problems, that stem from the fact that even the strongest of the theorems relating probability and frequency explicitly allow the possibility of probabilities that diverge from the frequencies. The logic of probability entails if something has the probability of 0.9, that does not mean it will occur. It does not even mean it will occur nine out of ten times. It means only that it will probably occur nine out of ten times, roughly, over the indefinitely long haul (or, perhaps, that it will probably occur nine out of ten times, in any large enough, not too carefully selected, set).[7]

Other modalized notions that resist reduction include causal relations, dispositions, capacities, and potencies. Among these, causal relations have a particular importance, and I will say some things specifically about them below. These are representative of the modal structures that appear in our scientific theories.[8] They are typically locally defined, but bear necessary connections to distributed structures or "patterns" in the manifold of categorical fact. We might say that they "encode information" about such patterns, and that information is drawn out in the kinds of empirical inferences in which they figure. A convenient way to picture things is that there is the ground level of categorical fact and then a second-order overlay of modal structures that play a role in epistemic and practical deliberation, or (as I will say), in guiding belief and decision.[9]

It is tempting to suppose that the structures on the second-order overlay are just re-descriptions of lower-level patterns, i.e., compact summaries of information about the pattern of actual fact. It is easy to see why we might find defined quantities that contain this kind of veiled information about distributed patterns useful. In general, we build a lot of useful, but extrinsic, information about how things generally hang together into our local representations of things, and this information can be unpacked to guide prediction and interaction with those things. When I describe a person as a wife, mother, and physician, for example, that is not an intrinsic description of her. It carries a lot of extrinsic information about the world, her place in it, and her habits. That information comes with a slew of expectations that are useful

[7] See Hajek (2011).

[8] I will focus on them. For discussion of causes, dispositions, and capacities, see Pearl (2000); Paul and Hall (2013); Kistler and Gnassounou (2007); Cartwright (1989).

[9] The image to have in mind is one of those old atlases in which some country or region is represented in colored ink on ordinary paper, followed by pages of transparencies that can be laid overtop of the original, adding information about anything from distribution of wealth to topology. In the analogy here, the categorical content is represented on the first pages as a four-dimensional mosaic of events. The modal content is drawn on the transparent overlays that assign chances to points, highlight causal pathways, give the dispositional profiles of the systems located in various regions, and so on. The structures on the overlay are second-order in that they contain information about patterns and relationships among the events that comprise the manifold.

in guiding interaction with her.[10] This is not just true of the way that we represent things in language. It has been known for a long time that the brain builds complex models of worldly regularities that guide expectation about what will happen, as well as offline predictions about what would happen if we acted in various ways.[11] Even perception is being increasingly understood as a process in which our brains do not just passively relay information but rather use an incoming signal as input to a stored schema that is used to predict the next signal before it comes in. The difference between the predicted signal and the one that occurs is then used to revise the schema (Clark 2013). The building of models that encode regularities that guide expectation (both about what *will* happen and what *would* happen under specified hypothetical conditions) is a common ground with representation in physics.

But as tempting as it is to see the chances, laws, and other modal structures built into our scientific models of the world as simply summaries of information about lower-level patterns, it turns out that no reduction is possible. These modal structures characteristically fail the logical test for identity with lower-level patterns for the reasons we saw already reflected above. Make any stipulation you like about what the laws are, and that stipulation will permit models in which there are exceptionless regularities that are not laws. There is a similar gap between claims about chance and claims about frequency. Make any stipulation you like about what the chances are, and that stipulation will permit models in which the chances diverge arbitrarily far from the frequencies. Something similar goes for dispositions, capacities, and causes. What this means is that claims about laws are not logically identical to claims about regularities, and claims about chances are not logically identical to claims about frequencies.

(Neo-)Humeans in the tradition of Lewis have tried to finesse this by combining stipulations about all of these into a single package and trying for a more holistic reduction.[12] But the difficulty remains. The problem is a generalization of the one

---

[10] To so much as call something a material object, or talk about its location in space, is to say something with very broad consequences about the possibility of various types of experience. It presupposes the whole embedding framework in terms of which material objects are defined. The constraints imposed by the embedding framework are the source of our most basic expectations about the world.

[11] The image of the brain as an engine of prediction can be found in various forms in contemporary neuroscience; see Bubic, von Cramon, and Schubotz (2010); Friston (2010); Helmholtz (1860); Kveraga, Ghuman, and Bar (2007). It remains an open question whether all brain processing can be subsumed by the predictive coding framework, but there is little doubt that the brain builds models that guide expectation, and that these models play a role in perception, imagination, and action.

[12] Lewis (1980) calls his view "Humean," and the terminology has become standard, though the view that goes under this label was almost certainly not Hume's own. Like Lewis, Hume denied that there are necessary connections between distinct existences, but he did not hold that beliefs about laws and chances reduce to beliefs about patterns in the manifold of categorical fact (Morris 2009; Strawson 1989; Beebee 2004). I have argued for a form of Humeanism closer (I believe) to Hume's own that cleaves to the denial of necessary connections,

that foils the more simple-minded reductions above; the logic of beliefs about laws-and-chances recognizes a modal gap between the facts about laws-and-chances and the categorical facts. Make any stipulation you like about what the laws + chances are, there are models of that law + chance package in which the categorical facts are very different than they actually are, and there are worlds in which the categorical facts are as they are, but the laws and chances are different. Let C be the categorical facts at a world W, and T be the Best Systematization of C. There are worlds at which C but not T (C is a model of other theories), and worlds at which T but not C (there are models of T in which not C). The persistent stumbling block for Humeanism is that there is a difference in truth conditions between structures on the second-order overlay and patterns in the manifold of categorical fact. These structures have a built-in inductive content in the form of implications for what would happen in hypothetical conditions that outruns any information about the pattern of actual fact. Information about the pattern of actual fact can provide evidence for claims about law or chance, but the relationship between the pattern of actual fact, on the one hand, and claims about the laws and chances, on the other, falls short of identity.

This can seem inconclusive for several reasons. A defender of reduction might say, that, since the difference in truth conditions can only be made out in modal terms, it is an empty difference. Or he might say that the claim that there are models of our law + chance packages in which the categorical facts are very different from what they actually are, and the claim that there are possible worlds in which the categorical facts are as they are, but the laws and chances are different, are both equivalent to the assertion that Humeanism is false. So the argument presupposes what it means to prove.[13] Or he might acknowledge that the Humean view closes a logical gap that our everyday notions of law and chance leave open, but bite the bullet. Barry Loewer—the most influential defender of (neo)-Humean reduction—adopts the last option. He acknowledges that the view is slightly revisionist about the everyday meaning of chance, but says (effectively): so much the worse for our everyday notion of chance (2004). This is a viable position only as long as the revised notion can serve all of the crucial functions of the everyday notion.

The problem with Loewer's response is that the revised notion *cannot* serve all of the crucial functions of the everyday notion. That brings us to the really telling objection against the Humean view which is that the Humean view has to close the logical gap between the modalized structures on the second-order overlay and

but drops the demand for reduction (Ismael 2015). I will follow the standard terminology here, using "Humean" to refer to the Lewisian tradition, without meaning to impute these views to Hume.

[13] The Humean can say this only at the cost of rejecting the identification of models of a physical theory with the physically possible worlds in which that theory holds.

categorical facts, but there is a difficulty with any attempt to do so that emerges when we adopt the side-on view and look at the function of these notions. No beliefs about mere description of patterns in the manifold of actual fact could play the role that beliefs involving these modalized notions play in guiding belief and action.[14] Creatures who need to make decisions about how to act have a need for beliefs whose inferential implications outrun beliefs about what merely happens, for they need to know what would happen if they acted in any number of ways, only one of which will be actualized. Beliefs about the results of hypothetical interventions in nature are indispensible in practical reasoning.[15]

3. FUNCTION

Let us start with a wide-angle view of why we construct models of the world at all. And here I do not just mean the relatively esoteric products of professional science; I mean the internal world-models that the brain constructs in order to help us navigate a complex and changing terrain. We are the only creatures who seem to have evolved to use models as the setting for an explicit form of practical reasoning (or at least do so with anything like the power and sophistication that we do; there is evidence that mice and other creatures do a rudimentary form of map keeping, but we have full-blown models of the world on which we represent ourselves and our ends). Our behavior (or, rather, our deliberate behavior, i.e., the willfully initiated movements of our limbs) is governed by a decision process that involves explicitly representing potential actions, imaginatively tracing out their effects, and making a choice about what to do based on projected outcomes. This process is our most powerful cognitive tool, one that gives us our primary advantage over other kinds of naturally evolved cognitive systems. Models provide the setting for this deliberative process. The added layer of representational mediation between stimulus and response gives us a kind of flexibility and foresight that holds perhaps our greatest advantage over natural competitors.

Functionally, constructing models is a human strategy for behavior management. Science is an extension of this basic strategy which involves the collectivization and systematization of information, the creation of models of varying scope, specially

[14] By "modalized notions," I mean notions that have a modal component to their content. I am arguing that modalized notions have different cognitive roles from their proposed non-modal counterparts, so even though the Humean wants to say that their descriptive content is exhausted by what they say about actuality, their modal implications are needed to capture their cognitive role. (If we assume that facts are as finely individuated as beliefs, we could run the argument for irreducibility with either facts or beliefs.) The sense in which these structures are second order is that they are functions of the first-order pattern of fact.

[15] The same is true, though less obviously so, for creatures who need to form expectations under ignorance, because such creatures need guides to belief that cover all epistemically possible situations, and for all such creatures the epistemically possible situations will include non-actual ones.

tailored for all different kinds of purposes. We make maps of outer space and build models of atoms, cells, and ecosystems. These models all play a role in our interaction with the natural environment. In this capacity, model construction is not merely a matter of copying. It involves restructuring, reorganizing, and reconfiguring information: integrating and reformatting it in ways that prepare it for use in inference or navigation. Models are tools. Their job is to facilitate interaction between an embodied agent and an open environment. Some of the structures defined in a model have the job of representing: tracking or mirroring localized elements in the landscape. In those cases, the account of how the models are used will support the kinds of localized correspondence that most people think of as paradigmatic of representation. We expect this kind of localized extensional correspondence, for example, between first-order elements in a model of space-time and localized events (e.g., a lightning strike or the decay of a radioactive atom). But that is a quite specialized function. There are also structures defined on our models that encode information about distributed features of the world like trends and currencies, the latest fashions, the value of the dollar, or the state of the union. And there are structures whose main function is to facilitate computation. We store information about dates and locations in formats that make it easy to compute duration and distance. And, in general, information will be encoded in different formats to facilitate different kinds of function. The lesson here is that models provide embedding frameworks for phenomena that package information for useful application in situ. This re-packaging can introduce a holistic restructuring that does not in general preserve piecemeal correspondence, and (more importantly for our purposes) introduces elements that do something other than simply reflect first-order features of the landscape.

Chances are easy to understand in these terms.[16] Chance is a species of statistical probability tailored to guide credence for creatures that have no direct source of information from the future. Statistical probabilities are objective, modalized quantities grounded in relative frequencies that guide expectation in open-ended classes of systems. They do not correspond to actual frequencies because actual frequencies can be skewed in a way that would make them unsuitable for that role. If a coin falls heads half the time, but all of those head-tosses occur before the birth of Socrates and after, say, 3011 CE, it would be stupid for you and I to take even odds on heads or tails. Chances reflect facts about stable relative frequencies over the short term in a way that is quite precisely designed to allow them to play their role-guiding expectation.

[16] Because of Lewis's influence, the problem that chance played in his metaphysics, and the pristine clarity of his own work on the subject, there is a very well-developed discussion of chance in the philosophical literature. These programmatic remarks about chance are supported in more precise detail in Ismael (2011b, 201). For some of the background on chance, see Bigelow, Collins, and Pargetter (1993) and Hall (1994).

The epistemic uses of models have to do with carrying information, computing, and predicting. But these are not the only uses. Models also guide our interactions with the systems they represent. In this manner, the ways in which we represent things will contain information that is useful for the purposes of intervention. The intervener does not simply need to know how things *are*; he needs to know how things *would* be if he acted on the world in various ways.[17] To think of models in purely epistemic terms is to forget about their practical role. To the embedded agent who doesn't just observe, but also intervenes in, his environment, the world is chock-full of opportunities and affordances. The terms in which he represents the world will be designed to disclose them. Causal relations are the generic form of these opportunities and affordances. Formally, causal relations are inductive generalizations of emergent relations among networks of variables that tell us what would happen to other variables in a network if we intervene on one. These relations are captured in DAGs (Directed Acyclic Graphs) that highlight strategic routes to bringing about ends.[18]

Recognizing the practical dimension of use is what we need in order to understand alethic modalities. Epistemic modality involves the notion of *how things might actually be, given what we already know*. Alethic modality involves the notion of *how things would be, under conditions that may or may not be actual*. It is the alethic modalities that have seemed to carry metaphysical commitments that have been uncomfortable to empiricists. This is because making out the modal content of an alethic modal claim involves quantification over specifically counterfactual (i.e., non-actual) possibilities. To say that A follows B as a matter of law, is to say that A *must* follow B, i.e., that A *could* not *fail* to follow B. To say that the association between A and B is not merely a correlation, but a cause, also adds some counterfactual force. It supports the inference that if one were (hypothetically) to bring about A, B would follow. In both cases, the extra modal force can only be made out in counterfactual terms. The modal force captures something crucial to the content of those notions. What does the modal force add? It does not add anything new to our beliefs about what *does* happen. But it does add something of practical importance that makes a difference to choice. You might try to bring about an exception to a regularity, but you would not want to try to bring about an exception to a law.

---

[17] The case of cause parallels that of chance. Just as in the case of chance, causes can be implicitly defined by their role in practical reasoning. And causes relate to correlations in a manner that is quite similar to the relationship between probabilities and frequencies. See Ismael (2012; 2016, chap. 5). For background, see Pearl (2000); Spirtes, Glymour, and Scheines (2000); Woodward (2003).

[18] One might think of causal beliefs as encoding implicit, conditional practical imperatives whose practical consequences are drawn out in deliberative application. The practical consequences are a little more complex than "do x." They say "do x if you want y to be the case," or "do x if you want y to be the case, and one of $\{z_1 \ldots z_n\}$ and none of $\{z^*_1 \ldots z^*_n\}$ obtain as well," . . . or something of this sort.

It would be a waste of time, i.e., a strategic mistake. To know that the relation between A and B is a causal one does not add anything to our stock of categorical beliefs; it signals that one could use the link strategically by manipulating A to bring B about.

Philosophers have focused on the counterfactual as the most basic alethic modality, but counterfactuals are just hypothetical statements with false antecedents. And in cognitive terms, the hypothetical statement is the more basic category. The role that beliefs about hypothetical circumstances play in practical reasoning is easy to discern.[19] When I am deciding how to act, I consider a range of actions. The way that I decide is by tracing out the downstream consequences of actions considered in the hypothetical. What would happen if I accept the Queen's Gambit or defend my knight? Should I take the beaten path or the road less traveled? The answers depend on what would happen if I did.[20] And there is no way of eliminating the modal content. Only some of the hypothetical futures I consider under the guise of potential actions will be actualized. The others are, and will remain, strictly counterfactual. One way of putting this is that epistemic modalities are to theoretical reason what alethic modalities are to practical reason. Looking back now, we can see more clearly why the attempts at reduction of laws to regularities, and chances to frequencies, failed. In both cases, the looseness of fit between the categorical facts and the structures that reside on the second-order overlay is essential to the function of those structures. Chances have the function of guiding expectation in open-ended classes of systems when we have general information about the distribution of values for some quantity in the class from which the system is drawn, but no specific information about the value the quantity takes in the case in question.[21] And the open-ended application means that chances have to range over possible, not merely actual, instances. They have to cover any system we might come across, and we have no way, in advance, of delimiting the ones we will come across from those we could. Claims about laws have specifically counterfactual implications because they have the function of guiding the kinds of purely hypothetical imaginings that are part of

[19] This insight is captured succinctly in Alison Gopnik's lovely dictum, "Counterfactuals about the past . . . seem to be the price we pay for counterfactuals about the future" (2009, 23). I would change this slightly to say that past counterfactuals are the price we pay for future *conditionals*. Counterfactuals and future conditionals are both species of hypothetical. Science deals generically with hypotheticals, and although hypotheticals give us the logical resources to define counterfactuals, it is the future conditionals that have the most basic cognitive function.

[20] On the logic of these imaginative explorations and what distinguishes them from purely epistemic reasoning, see Joyce (2002); Anscombe (1963); Ismael (2011a).

[21] There are well-defined probabilities only when there are stable relative frequencies across arbitrary subselections from the class. If the class does not have the right structure, or we have specific information about the instance in question, then chances are not relevant in the same way.

deliberation. To play this role, these quantities have to have implications that guide belief about hypothetical, potential futures.[22]

What do we say about these structures, then, if they do not describe what happens? We say that they are inductions on patterns in the manifold of facts that supply us with best-guesses-under-the-circumstances for what will happen, and also about what would happen if we acted in various ways. The regularities that underwrite the modalized structures that embody these best guesses are part of the pattern of actual events. The modal force is an inductive projection of those patterns into the unknown and the purely hypothetical. Naturalistic philosophers looking for a complete, non-redundant catalogue of the basic objects, quantities, and relations of which the world is composed can look to the categorical part of physics. But science is not just about reflecting what is the case. It is also charged with providing representations that can function as a convenient user interface for creatures with our combination of limitations and needs. Overcoming the limitations that our native equipment imposes on how far we can see, and how effectively we can intervene, sets the task for science (and, indeed, for cognition more generally). Scientific models—on the local and global scale—are embodiments of our very best inductive practices. I am suggesting that the modal content of our models—the overlay of laws, dispositions, capacities, and potencies—are to be understood in terms of their role guiding prediction and decision.[23]

4. MODELS AS PPS'S TO FEP'S

Scientific induction is a holistic enterprise that proceeds in two steps. The first step involves the use of information about local matters of particular fact to produce models that contain locally defined modal structures which act as guides in belief and decision.[24] The second step draws out modal implications that are built into the content of models at the first stage. Edwin Hutchins, in an account of the cognitive ecology of a naval vessel, uses a phrase to describe navigational instruments

---

[22] This is not their only role. We also care about what would have happened in the past if we had acted differently, even though there is no possibility now of changing that fact, for assigning responsibility, and learning practical lessons, for example.

[23] And if asked what the specifically modal content represents/stands-for/corresponds to, I say either (using "represents" in a deflationary way) that it represents modal facts, or (using "represents" in a non-deflationary way) that it does not represent anything. The ambiguity between inflationary and deflationary conceptions makes the vocabulary of representation famously fraught. I have tried to be explicit in the text where I mean it in an inflationary sense to avoid confusion. See Price (2011) and Thomasson (2015) for discussion of the deflationary alternative. In either case, I deny that either reduction or reification is needed for realism about these structures.

[24] "Locally defined," here, means only assigned to space-time points or regions. It does not mean "describes properties intrinsic to those regions."

like the compass and alidade that is quite helpful here. He calls them "partially prepared solutions to frequently encountered problems."[25] That phrase captures quite precisely the functional status of the modal quantities introduced into our models in the first stage of theorizing. We use models to predict, compute, and intervene. The structures that form the modalized overlay are designed to facilitate those tasks. They encode partial solutions to the kinds of problems that situated agents frequently face. Viewing models as PPS's to FEP's sets the agenda for interpretation of a wide class of structures that are important in science. I have indicated summarily how work already done on chances and causes fits neatly into this mold. Laws, dispositions, and capacities are differentiated from these structures, and also from one another, by their categorical content and modal implications. They might also be viewed as encoding inductive inferences that can be drawn on by the situated agent under a variety of conditions to guide belief and action.

5. LEWIS'S PROGRAM

The title of this paper is intended to evoke David Lewis's (1980) "A Subjectivist's Guide to Objective Chance," which sought to ground notions of objective chance in subjective probabilities. In that article, Lewis set about trying to show how someone who was a subjectivist about probability—i.e., someone who thought that probabilities expressed subjective degrees of belief—might also recognize objective chance. In his words:

> We subjectivists conceive of probability as the measure of reasonable partial belief. But we need not make war against other conceptions of probability, declaring that where subjective credence leaves off, there nonsense begins. Along with subjective credence we should believe also in objective chance. The practice and the analysis of science require both concepts. Neither can replace the other.[26]

Lewis's strategy was to first identify a use that creatures like us would have for beliefs about chance, which he embodied in his Principal Principle. He then treated the Principal Principle as an implicit definition of chance, and looked for something to assign as extension to beliefs about chances. He looked, that is to say, for something for those beliefs to represent. I want to pause to relate the view that I have been offering to Lewis', because although I depart from Lewis in the details, I am effectively doing for alethic modality what his paper set out to do for epistemic

[25] The phrase occurs in Hutchins (1995, 134). He remarks that the phrase also has currency in computer science.
[26] Lewis (1980, 267).

modality. I am suggesting thatwe do not need to reify modal fabric of nature in order for it to be objective. We can borro Lewis: defining chance functionally in terms of its role guiding ing chance as the objective face of credence. We generalize those insi lows. First, we define all of the modal structures generated by science (chan causes, dispositions, laws) functionally as PPS's to FEP's, understanding the alethic modalities in terms of their role in guiding decision rather than belief.[27] Second, we see these structures as the objective faces of structures that describe the embedded agents' epistemic and practical relations to events.[28]

Lewis offers his Best Systems Analysis (BSA) as a content-preserving reduction of beliefs about laws and chances to beliefs about patterns in the Humean mosaic. We hold that laws and chances and other intermediate structures embody the inductive outputs of a theoretical process that is more or less loosely modeled by the BSA. But we modify Lewis's account in two ways. First, as Lewis describes it, the BSA is an abstract logical recipe for forming beliefs about laws and chances from a full description of the manifold of categorical fact. We modify it to look more like real science, embedding it in the epistemic context that gives science its purpose. Beliefs about laws and chances are now seen as embodying inductive hypotheses that tell us how to project perceived patterns in a partially known mosaic in order to guide both expectations and interventions. Experimental practice, which has little place in the Lewisian recipe, is now center stage. This is a subtle matter that I haven't argued here, but the reason that experimental practice becomes important is that the experimental side of science is crucial to establishing its alethic content. Second, we hold that the BSA is a recipe for forming beliefs about laws and chances, but not a content-preserving reduction, because we hold that laws and chances have modal implications that outrun beliefs

---

27 The take-home lesson is that reifying these quantities in the fabric of the world is not the right view. The main argument against reification is that the interpreter who reifies modality in the absolute fabric of Being has to square the *semantic* content with the *practical* function. If he makes beliefs about counterfactual possibilities beliefs about other worlds, he has to explain why beliefs about such things would guide action in our world. If he makes them beliefs about linguistic entities of some kind, he gives them something actual to refer to, but then he has to explain why beliefs about linguistic entities would guide action. And if the answer is that they do so because they represent possibilities, he has gotten nowhere.

28 There seems to be a confused idea that if chances and laws are not reified, they are subjective in an objectionable way, i.e., not "out there" in the world as proper objects of scientific study. That is just a mistake. They are inductions grounded in what is "out there" crafted to provide solutions to problems introduced by the agent's perspective. Lewis's "A Subjectivist's Guide to Objective Chance" got the senses in which these structures are objective right. Chances are not descriptive of anyone's degrees of belief. They are rather structures meant to guide degrees of belief. They are objective in two senses: (i) in the sense that they are based on an inductive procedure that takes its departure from objective facts about the pattern of events, and (ii) in the sense that they permit a distinction between what the chances are and what I think the chances are.

t what actually happens. We make a distinction between the content of beliefs out laws and chances and the facts that provide the epistemic grounds, or support, those beliefs. We want to say that PPS's to FEP's are inductions *grounded epistemically in* objective facts about patterns in the Humean mosaic, but they have modal implications that *outrun* those patterns. Those modal implications, moreover, play crucial roles in the epistemic and practical life of the embedded agent. Lewis assumes that all beliefs have the function of representing/standing for/or reflecting some aspect of what there is, and so he looks for elements or patterns in the Humean mosaic for beliefs about chance and laws to represent. We agree with Lewis that the world is just a vast mosaic of local matters of particular fact. We depart from Lewis in denying that all belief has the function of representing/standing for/or reflecting some aspect of what there is.[29] Believing and cognizing are natural activities that are part of a behavior management strategy that involves anticipating what will happen next and acting to forestall, avoid, bring about, or promote outcomes. We hold that laws, chances, causes, and other modal outputs of science play a role in this process.

6. MODALITY FOR THE EMPIRICIST

The result is a form of empiricist realism about modality that steers a path between reduction and reification. The pressure to reify comes partly from an argument that goes like this: if intermediate structures do not represent patterns in the manifold of actual fact, then what do they represent? And we are given two options: either they represent something non-categorical or they do not represent anything at all.[30] PPS's to FEP's do not fit clearly into either of these categories. Part of the point of the discussion here was to break down this simple dichotomy. There is, in my view, no more compact story in principle that relates beliefs about modal structures to categorical facts than one that says how these beliefs are formed and the role they play in our cognitive and epistemic lives.

Is this simply instrumentalism? Yes and no. It does hold that modal structures are instrumental in the sense that they represent the world in a form that is poised to play a role in practical and epistemic inference. But it is not the kind of instrumentalism sometimes associated with Duhem according to which instrumental structures

---

[29] For further discussion, see Ismael (2014, 2015).

[30] Here is a characteristic statement of the options (reading "categorical" for "non-modal") due to Shalkowski (1994, 670): "If modality is grounded in reality, it is either a primitive or a non-primitive feature of that reality. If it is primitive, then there is nothing non-modal in virtue of which reality possesses modal characteristics—there are no non-modal facts that wholly constitute modal facts. If it is not primitive, then there is something non-modal in virtue of which modality is present in reality—there are non-modal facts that wholly constitute modal facts."

are empty nodes in a formal calculus that do not have any representational significance of their own. It is much closer to the instrumentalism of Dewey, according to which all belief is both contentful and geared toward action (Godfrey-Smith 2010). It does not matter much for my purposes what philosophical vocabulary is used to describe this, except that we should refuse to say that this account entails that there are no such things as laws or chances. We should say that the account that we have given tells us what laws and chances *are*. And we should add that once we understand what laws and chances are, we will see that forming beliefs or making claims about laws and chances does not commit us to the existence of possible worlds, or any other kind of ontologically substantive posit. Everything that there is to know about laws, chances, and other scientific modalities is given in the account of how beliefs about chances are formed, their inferential implications, and the role they play in our practical and epistemic lives. The philosophical vocabulary that I prefer to describe the account is deflationism. Deflationism holds that there is a substantive story about the formation of beliefs about laws and chances, and the role they play in our epistemic and practical reasoning. But the laws and chances themselves are just, so to speak, shadows of law and chance beliefs.[31]

It should be acknowledged, however, that there are other sources of the pressure to reify. One such pressure comes from the basic non-Humean intuition that laws act as the iron enforcers of regularity in nature. I have said nothing to address this intuition here. It has to be addressed by providing an alternative account of what makes our world hospitable to inductive practices. That is a rather different topic. Here I have been mostly focused on providing a non-reductive empiricist account of modal belief.

7. TO SUM UP

I have argued that claims about laws, chances, and other modalized structures encode inductions on observed regularities in a form that is tailored to solve the kinds of practical and epistemic problems that beings like us—beings with limited sources of information about the world that gather and store information and use it to guide behavior—face. The account of how we form beliefs about such structures is given by a description of scientific practice, modeled loosely on Lewis's Best Systems Analysis.

[31] The metaphilosophical framework has been developed by a number of people. See, in particular, Price (2011) and Thomasson (2015). In this framework, correspondence, representation, and truth are all deflated. Although the availability of deflationism deflects pressure to reify that comes from the felt need to provide truthmakers for modal beliefs, there is a pressure to reify that appeals to the basic non-Humean intuition that the regularity in nature is inexplicable without modal connections among events. I take that pressure seriously, but it is a battle to be fought on other grounds.

I denied that such beliefs are reducible to beliefs about categorical facts. The reason was that they have modal implications that are crucial to their function and that outrun any claim about what actually happens. I also denied that the extra content does anything more than to project observed regularities into hypothetical situations of the kind that we entertain imaginatively in the course of decision. I characterized the function of these structures on our models generically by saying that they act as PPS's to FEP's. Individually, their function is given by a detailed description of the particular role they play in epistemic and practical reasoning. Chances, for example, provide best guesses for creatures confronting an unknown future. Causes highlight strategic routes for bringing about ends. Laws encode general inductive hypotheses that constrain both action and belief. Dispositions and capacities encode other kinds of useful inductions that guide the interaction of the embedded agent with the systems to which they are ascribed. I have not tried to talk anyone out of a metaphysically inflated notion of modality; I have just tried to suggest that there is a sensible story for empiricists to tell about the modal commitments of science that does not saddle them with an objectionable metaphysics. This opens up space for an empiricist account of the central modalized concepts of science that looks at the inductive content they encode, and the role that content plays in our practical and epistemic lives, without trying to reduce or eliminate it.

REFERENCES

Anscombe, G. E. M. 1963. *Intention*. 2nd ed. Oxford: Blackwell.

Beebee, Helen. 2004. *Hume on Causation*. London: Routledge.

Bigelow, John, John Collins, and Robert Pargetter. 1993. "The Big Bad Bug: What Are the Humean's Chances?" *British Journal for the Philosophy of Science* 44 (3): 443–62.

Bubic, A., D. Y. von Cramon, and R. I. Schubotz. 2010. "Prediction, Cognition and the Brain." *Frontiers in Human Neuroscience* 4 (25): 1–15.

Cartwright, Nancy. 1989. *Nature's Capacities and Their Measurement*. Oxford: Clarendon Press.

Clark, Andy. 2013. "Whatever Next? Predictive Brains, Situated Agents, and the Future of Cognitive Science." *Behavioral and Brain Sciences* 36 (3): 181–204.

Friston, Karl. 2010. "The Free-Energy Principle: A Unified Brain Theory?" *Nature Reviews: Neuroscience* 11 (2): 127–38.

Godfrey-Smith, Peter. 2010. "Dewey and the Subject Matter of Science." In *Dewey's Enduring Impact: Essays on America's Philosopher*, edited by John Shook and Paul Kurtz, 73–86. Amherst: Prometheus Books.

Gopnik, Alison. 2009. *The Philosophical Baby: What Children's Minds Tell Us About Truth, Love, and the Meaning of Life*. New York: Farrar, Straus and Giroux.

Hájek, Alan. 2011. "Interpretations of Probability." In *Stanford Encyclopedia of Philosophy*. Stanford University, 1997. First published October 21, 2002; substantive revision December 19, 2011. http://plato.stanford.edu/archives/win2012/entries/probability-interpret/.

Hall, Ned. 1994. "Correcting the Guide to Objective Chance." *Mind* 103 (412): 505–18.

Helmholtz, H. (1860) 1962. *Handbuch der physiologischen Optik*. Vol. 3. Translated by J. P. C. Southall. New York: Dover.

Hutchins, Edwin. 1995. *Cognition in the Wild*. Cambridge, MA: MIT Press.

Ismael, Jenann. 2011a. "Decision and the Open Future." In *The Future of the Philosophy of Time*, edited by Adrian Bardon, 149–68. New York: Routledge.

Ismael, Jenann. 2011b. "A Modest Proposal About Chance." *Journal of Philosophy* 108 (8): 416–42.

Ismael, Jenann. 2012. "Causation, Free Will, and Naturalism." In *Scientific Metaphysics*, edited by H. Kincaid, J. Ladyman, and D. Ross, 208–36. Oxford: Oxford University Press.

Ismael, Jenann. 2014. "Metaphysics on the Sydney Plan." In *Philosophical Methodology: The Armchair or the Laboratory?*, edited by Matthew C. Haug, 86–103. New York: Routledge.

Ismael, Jenann. 2015. "How to Be Humean." In *A Companion to David Lewis*, edited by Barry Loewer and Johnathan Schaffer, 188–205. Oxford: Wiley Blackwell.

Ismael, Jenann. 2016. *How Physics Makes Us Free*. New York: Oxford University Press.

Joyce, James. 2002. "Levi on Causal Decision Theory and the Possibility of Predicting One's Own Actions." *Philosophical Studies* 110: 69–102.

Kistler, Max, and Bruno Gnassounou, eds. 2007. *Dispositions and Causal Powers*. New York: Ashgate.

Kment, Boris. 2014. *Modality and Explanatory Reasoning*. Oxford: Oxford University Press.

Kveraga, K., A. S. Ghuman, and M. Bar. 2007. "Top-Down Predictions in the Cognitive Brain." *Brain and Cognition* 65: 145–68.

Ladyman, James. 2004. "Constructive Empiricism and Modal Metaphysics: A Reply to Monton and Van Fraassen." *British Journal for the Philosophy of Science* 55 (4): 755–65.

Lewis, David. 1980. "A Subjectivist's Guide to Objective Chance." In *Studies in Inductive Logic and Probability*, edited by Richard C. Jeffrey, 2:263–93. Berkeley: University of California Press.

Loewer, Barry. 2004. "David Lewis's Humean Theory of Objective Chance." *Philosophy of Science* 71 (5): 1115–25.

Morris, William Edward. 2009. "David Hume." In *Stanford Encyclopedia of Philosophy*. Stanford University, 1997–. First published February 26, 2001; substantive revision May 15, 2009. http://plato.stanford.edu/archives/spr2013/entries/hume/.

Paul, L. A., and Ned Hall. 2013. *Causation: A User's Guide*. Oxford: Oxford University Press.

Pearl, Judea. 2000. *Causality: Models, Reasoning, and Inference*. Cambridge: Cambridge University Press.

Price, Huw. 2011. *Naturalism without Mirrors*. New York: Oxford University Press.

Shalkowski, Scott A. 1994. "The Ontological Ground of the Alethic Modality." *Philosophical Review* 103 (4): 669–88.

Spirtes, Peter, Clark Glymour, and Richard Scheines. 2000. *Causation, Prediction and Search*. 2nd ed. Cambridge, MA: MIT Press.

Strawson, Galen. 1989. *Secret Connexion: Causation, Realism and David Hume*. Oxford: Oxford University Press.

Thomasson, Amie L. 2015. *Ontology Made Easy*. New York: Oxford University Press.

Van Fraassen, Bas. 1980. *The Scientific Image*. Cambridge: Cambridge University Press.

Woodward, James. 2003. *Making Things Happen*. New York: Oxford University Press.

6

So Long, and Thanks for All the Fish

METAPHYSICS AND THE PHILOSOPHY OF SCIENCE

P. Kyle Stanford

THE ADMIRABLY IMMODEST goal of metaphysical inquiry has always been to answer our deepest questions concerning the fundamental constitution, organization, and character of the world, and the traditional methods of that inquiry in the analytic tradition have been clear-headed analysis of and reflection upon the concepts and linguistic categories we use to engage that world.* But at least since the time of the Scientific Revolution, metaphysics has often seemed to be in something of an intellectual free-fall, with history bearing witness to the sequential collapse of progressively weaker and weaker rationales for thinking that the traditional methods of metaphysical inquiry have any claim to inform us about the world's fundamental constitution, organization, or character. Descartes, for example, thought he had decisive reasons for believing his own concepts to be the creations of an omnipotent, omniscient, and benevolent God who would (therefore?) not allow these concepts to deceive him so long as he was careful to reason from and about them correctly. But as modern scientific knowledge accumulated, it ultimately came to seem imprudent, to say the very least, to rely on any such theistic guarantee of fidelity for our

* And thanks to Rob Wilson for the idea of repurposing Douglas Adams's (1984) clever title in this way, and to Michael Strevens, Katherine Brading, Jeff Barrett, Penelope Maddy, L. A. Paul, Eleanor Knox, Alexander Reutlinger, JackJack Ritchie, Collin Rice, Craig Callender, James Ladyman, Ken Waters, Katie Elliott, audiences at the Metaphysics and Philosophy of Science conference in Toronto and Universidad Autónoma Metropolitana in Mexico City, as well as others I have inexcusably forgotten for useful input on the subject matter of the paper itself.

conceptual apparatus. It no longer seems a safe bet that our concepts have been bequeathed to us by a divine creator, and even if they have, it does not appear that enabling us to attain a clear and accurate understanding of the world's fundamental constitution and principles of operation was high on Her list of priorities.

But even as the advance of science called any such divine guarantee of fidelity into serious question, the pioneering work of Immanuel Kant seemed to offer an appealing rationale for *something* like traditional metaphysical inquiry by means of his Copernican Revolution. On this Kantian rationale, an investigation of fundamental concepts like cause, object, and property tells us what the world must be like in its most fundamental constitution simply because those concepts are the ones we use in constructing genuine *experience of a world* out of the raw materials that are available to us in sensation and reflection. Our most fundamental concepts, Kant suggests, do not arise in us *in response to* experience but are instead what we use to understand what we encounter in sensation *as* experience of a single, coherent, intelligible world at all. Their informativeness is guaranteed by the distinctive role they play in *constituting* the world as we experience it, rather than because we think some outside agent takes an interest in ensuring their fidelity.

But subsequent scientific developments have not been much kinder to this Kantian rationale for metaphysical inquiry than to Descartes's theistic one. First, as the American pragmatists insisted, even the most fundamental conceptual categories we have now appear to be considerably more plastic than Kant allowed. After all, Kant's famous Table of Categories is ultimately generated from a corresponding table of logical judgments, whose claim to universality was rooted in the fact that logic had not changed substantially in the two thousand years since Aristotle; sadly for Kant, he made this argument a mere hundred years before Frege's *Begriffsschrift* inaugurated modern logic with a radical transformation of the Aristotelian approach whose presumed universality and permanence was the ground on which Kant thought he could identify metaphysical fixed points in the world-constituting conceptual apparatus of human beings. Even more importantly, scientific developments seem to have illustrated more clearly than any others just how wide a gap there is between the world as we may be constrained to *experience* it and the world as we might reasonably hope to *conceive* of it by means of our scientific theories. That is, even if Kant was right to suggest that our experience must be experience of a world whose geometry is Euclidean, say, or whose causal connections must be presumed to be deterministic, we seem to have developed scientific theories that do not respect these constraints but are nonetheless substantially intelligible to us as candidate descriptions of the fundamental constitution of nature. None of this implies or even suggests that there *are no* constraints on the forms of possible human experience or even on the ways in which it is open to us to conceive of the fundamental constitution of the world (in fact, that suggestion seems implausible in light of what else we think we know about ourselves), but it is to insist that (1) we have learned to recognize a deeper division than Kant saw between constraints on our possible experience of the world and constraints on the ways in which it is possible for us to conceive of that world theoretically; (2) we do not seem able to gain access to either sort of constraint in the ways that Kant thought were open to us; and (3) even when we can identify limits on our ability to *conceptualize* rather than experience the world, it is not at all clear which of these are necessary consequences of the human conceptual apparatus as such, rather than contingent artifacts of a particular historical, cultural, or otherwise locally restricted perspective. I freely confess that I am able to make no concrete sense of the idea that we might come to conceive of the world in a way that does not recognize the existence of objects with properties, but scientific history leaves me unwilling to infer that this conceptual limitation any more represents an obligatory feature of any possible human (or "discursive") theoretical conception of the world than did the respective insistences of earlier generations that the idea of non-Euclidean space, or action at a distance, or wavelike propagation without a physical medium of transmission, or any purely mechanical and/or non-vitalistic conception of living organisms, were themselves literally nonsensical. The rug thus seems to have been pulled out from under Kant's Copernican guarantee of universal applicability for the fundamental metaphysical categories he identified, and if it *is* possible for us to discover universal or obligate features of the human conceptual apparatus as such, empirical psychology now seems a much better bet for their identification than introspection concerning what presently makes sense to us, what presuppositions presently seem required in order for us to have experience in the way that we do, or what conceptual possibilities seem presently open or closed to human beings who happen to share the pretty rarified air of early twenty-first-century academic philosophy.

Even as this Kantian strategy has faded somewhat unquietly from view, however, analytic philosophy has seemed to promise yet another, though still weaker, Copernican rationale and method for metaphysics. At least since the linguistic turn, some philosophers have seemed tempted to insist that conceptual analysis tells us about the fundamental constitution of the world because it is only by means of our concepts that we engage that world—it is then, in some sense, our concepts and the associated meanings we assign to our terms that determine what we are prepared to *count* as a "physical object," an "injustice," or a "peanut butter sandwich." But bereft of anything like the Cartesian or Kantian rationales for the fidelity or even the fixity of such concepts, this would seem a very cheap sense in which our ideas or language determine anything about what the world itself is like. It seems much more natural and less obscure to note that such conceptual analysis at best tells us something about how we (presently) think and talk about the world, and not about the characteristics of the entities and events that make it up. And the lasting *significance* of even this much more modest Copernican inquiry is no less compromised than Kant's own version by our pervasive evidence of dramatic interpersonal variation and historical evolution in even the most fundamental conceptual resources available to particular agents in particular times and places. Conceptual analysis thus seems to emerge as a better vehicle for exploring the distinctive features and limits of a particular, local conceptual scheme or cultural inheritance than for identifying ways that the world must be insofar as human beings as such are able to understand, constitute, or conceptualize it at all.

It might seem, then, that the advancement of science has been downright dangerous if not lethal to the notion that there are legitimate and distinctively metaphysical forms or methods of inquiry by which we may acquire substantive knowledge about the fundamental constitution, organization, or character of the world. Perhaps it is small wonder that in the headiest days of logical positivism and logical empiricism, many philosophers of science saw metaphysics as a sufficiently bankrupt and destitute enterprise that they confidently declared an indefinite moratorium on it, famously embodied in the tradition of members of the Vienna Circle loudly declaring "M" whenever they judged one of their fellows to be seeking to articulate a metaphysical proposition (a tradition still occasionally honored in my own department, though with considerably more irony and good humor than I suspect was the case in Vienna in the 1930s).

With logical positivism and logical empiricism long out of philosophical fashion, however, metaphysics seems to be making a roaring comeback in many quarters, and not only among those analytic philosophers who are most proudly ignorant and/or dismissive of scientific developments, but also and more importantly among many of those who are most intimately aware of and profoundly interested in the details of our best contemporary scientific theories. At least some philosophers seem to have decided it was not science but instead an excessively narrow philosophy of science that brought metaphysics into ill repute, and that our best scientific theories actually constitute a vast and underexplored resource or partner for metaphysical inquiry. But there are a variety of importantly different ways in which science seems to be playing midwife to this metaphysical renaissance, and we would do well to distinguish them from one another.

The first project in this new wave is something we might call "scientific metaphysics" or perhaps even "scientistic metaphysics." It is perhaps best conceived as preserving the traditional *aim* of metaphysical inquiry to inform us about the fundamental constitution of the world while insisting that consulting our best scientific theories is the principal means by which this aim may be successfully accomplished. Scientistic metaphysicians characteristically start with our best scientific theories and then *use* them to try to answer questions like whether time has a privileged direction, whether organisms have essential properties, and whether the universe is most fundamentally constituted out of structure itself, rather than consisting of objects with properties.

Of course, scientistic metaphysicians rarely hold the simplistic view that we can simply read metaphysical truths off of the claims of our best science (what would be left for metaphysicians to do?), but instead typically subscribe to the more broadly naturalistic notion that it is by consulting our best scientific theories in *some way or other* that we can and should satisfy our metaphysical curiosities. In their recently influential contribution to this enterprise, for example, James Ladyman and Don Ross suggest that "one metaphysical proposal . . . is to be preferred to another to the extent that the first unifies more of current science in a more enlightening way" (2007, 66), but they are also quick to point out that we must take note of how our best theories are "practically put to work" (119) in making this determination. Most fundamentally, Ladyman and Ross insist that "the *raison d'être* of a useful metaphysics is to show how the separately developed and justified pieces of science (at a given time) can be fitted together to compose a unified world-view" (45) and that "any new metaphysical claim that is to be taken seriously should be motivated by, and only by, the service it would perform, if true, in showing how two or more specific scientific hypotheses jointly explain more than the sum of what is explained by the two hypotheses taken separately, where a 'scientific hypothesis' is understood as an hypothesis that is taken seriously by institutionally *bona fide* current science" (30). It thus seems that we must take Ladyman and Ross at their word when they note that by the very term 'metaphysics' they will "refer to the articulation of a unified world-view *derived from* the details of scientific research" (65; my emphasis).

The first thing to notice about this project of scientistic metaphysics is that it simply cannot satisfy the traditional aims of metaphysical inquiry without the aid of substantial further assumptions that are themselves at least contentious if not dubious. Perhaps most importantly, if we wish to see our best scientific theories as giving us answers to the traditional concerns of metaphysics, then it seems we must first embrace a quite strong version of scientific realism. That is, we can only be as confident in our answers to metaphysical inquiries as we are in the truth (and completeness) of the science from which they are derived (in whatever way). So the more confidence we have that the long-distant future of our own scientific theorizing will reflect revolutions, additions, and/or emendations as profound as those that separated Einstein's physics from Newton's, or contemporary thermodynamics from the caloric fluid theory of heat, or Mendel's theory of inheritance from Weismann's, the less confident we can be in the answers we get to our metaphysical inquiries by consulting (or unifying, or whatever) those contemporary scientific theories. And our confidence that profound alterations of some sort are still to come might be considerably enhanced by the recognition that the two best-confirmed physical theories we have (special relativity and quantum mechanics) are logically inconsistent with one another in a perfectly straightforward and fundamental way (see Barrett 2003).

Moreover, it appears that scientistic metaphysicians are asking us to invest an improvident faith in precisely those parts or aspects of our best scientific theories that seem *most* vulnerable to periodic wholesale transformations.[1] After all, recently influential versions of scientific realism have modestly proposed that it is only the sort of formal *structure* that we find preserved between Fresnel's and later versions of the wave theory of light (and that persists into contemporary accounts of electromagnetic radiation) that is reliably preserved in all successors of (sufficiently) successful theories. Alternatively, other realists have even more modestly suggested that we can at least retain our confidence in the existence of *entities* playing one or more distinctive roles (e.g., being deployed in a complex and fine-grained way to investigate properties of *other* entities) in our theorizing about the world.[2] But especially within physics (the science from which scientistic metaphysicians seem to most frequently seek metaphysical guidance), it seems quite hard to make the case for metaphysical continuity itself across successive versions of successful theories, no matter how high we set the relevant standard of success. Was there ever a scientific theory more successful in its day than Newton's mechanics? But since the early twentieth century, our fundamental ontology has expanded to include fields, wave/particle duality, and much else that Newton never even so much as recognized as among the ontological possibilities. Thus, however hard the history of science and/or other argumentative considerations make it to sleep at night as some variety of scientific realist, I think they should make it considerably harder for scientistic metaphysicians to do so.

To be sure, arguments can and have been made for adopting forms of scientific realism strong enough to justify treating our most successful theories as metaphysical oracles in this way, especially when we compare the fruits of science to those of traditional metaphysical inquiry: science has helped us to achieve vast new abilities of prediction and control over nature, giving us cell phones, antibiotics, and space travel, while philosophy has given us mostly various sorts of headaches, and science has made real progress on answering its central questions and problems whereas philosophy has not obviously made any. Most generally (as I think the ghost of Groucho Marx whispers late at night in the ears of many philosophers of

[1] My thanks to Michael Strevens for pointing out that this line of argument is distinct from that of the previous paragraph.

[2] The former suggestion has been advanced most influentially by John Worrall (e.g., 1989) and the latter by Ian Hacking (e.g., 1983).

science[3]), the almost literally incredible practical accomplishments of the scientific enterprise can sometimes make philosophical inquiry feel a bit pale or even silly by comparison. But instead of therefore granting the sort of oracular privilege once reserved for first philosophy to our best scientific theories instead, I think a more appealing naturalistic response is simply to deny that any of our beliefs at all should enjoy such a privilege. The heart of the most appealing and plausible form of philosophical naturalism, I suggest (see Stanford, forthcoming a), is the idea that there is only a single, integrated project of inquiry into the world and our own place within it, a project that philosophers can and should pursue shoulder-to-shoulder with scientists themselves. And such a project cannot legitimately single out *at the beginning of that inquiry* just one part (e.g., "institutionally *bona fide* science") of what Quine famously called "the inherited world theory as a going concern" (1981, 72) and treat it as automatically making inviolable or even just privileged claims to knowledge. Our picture of the world and our own place in it is responsible for making good sense of *all* the evidence we have simultaneously, and this must include historical evidence of the profound metaphysical discontinuities between successive generations of successful scientific theories just as surely as it must include the undoubtedly impressive practical achievements of those same theories.

This conception of naturalized inquiry, however, would not alone suffice to rule out a quite different form of scientifically-inflected metaphysical revival that we might call "complementary metaphysics," recently pursued influentially by philosophers like L. A. Paul. In a recent article (2012), Paul breaks with the earlier tradition of metaphysical inquiry I describe above by explicitly and enthusiastically endorsing the idea that there are no special and distinctive philosophical *methods* for conducting metaphysical inquiry, holding instead that the methods of metaphysics (such as inference to the best explanation or the use of thought experiments) are simply a subset of those used within science itself,[4] but she insists that metaphysics nonetheless

[3] Marx famously resigned from the Friar's Club of Beverly Hills with a telegram that read, "Please accept my resignation. I don't want to belong to any club that will accept people like me as a member."

[4] Even this claim seems to me to involve considerable sleight of hand, as the appeal to inference to the best explanation ignores the fact that in scientific contexts it is typically what Peter Lipton (1991) describes as "likeliness" that is most centrally at issue in deciding on the "best" explanation, while in metaphysical contexts such judgments must rely on what Lipton calls "loveliness" instead (see also Saatsi's chapter in this volume). Also, the further features Paul insists are common to theory evaluation in both science and metaphysics (e.g., choosing among empirical equivalents on grounds of simplicity or other pragmatic virtues) are precisely those that generate concerns about whether all theoretical desiderata in science are actually truth tracking. The most natural exception would be explanatory power, but my discussion of the examples that follow in the main text is intended to suggest that the explanatory power asserted for competing metaphysical proposals of the sort that Paul offers is more apparent than real.

enjoys a special and distinctive *subject matter* because "metaphysics involves questions about features of the world that are prior to those described by science" (5).

On Paul's account, "metaphysics concerns the search for general and fundamental truths about the world," and "[t]he metaphysician engaging in such a search wants to determine the natures of the world, especially the fundamental natures of the world" (2012, 4). Lest this seem simply a dramatic way to describe the aims of scientific investigation itself, Paul argues that this distinctively metaphysical inquiry remains importantly distinct:

> [T]he different approaches are not in tension, for the ontological account involves features of the world that are metaphysically prior to those of the scientific account. The ontological account describes the metaphysically prior categories and constituents of the physically fundamental entities, and in this sense describes features of the world that are more fundamental than those of the natural sciences. (5)

For those who might be suspicious of such "metaphysically prior categories" or features of the world supposedly "more fundamental" than those described by natural science, Paul offers a number of specific examples of questions that the sort of inquiry she has in mind seeks to address:

> For example, when a fundamental physics takes fields to be the most physically fundamental entities of the world, an ontology will take the theory of the world a step further, by describing the ontological categories of the fundamental constituents of these fields. The substrate-attribute theorist will take fields and particles to be substances with properties, so will take both fields and particles to be substratum-attribute constructions from members of the fundamental categories of substance and attribute, while the bundle theorist will take fields to be bundles of property intensities and location properties, which are members of the fundamental category of property. And so on—the ontological account will take the properties postulated by the scientific theory, such as properties of charge, spin, and mass, and ask if they are fundamentally universals or fundamentally tropes (or fundamentally something else). (5)

The most pressing worry here, I think, is whether these metaphysical proposals are really adding anything to the conception of the fundamental constitution of the natural world offered by the relevant scientific theories themselves. I suspect that there is no more than what Ernest Nagel memorably called a "merely verbal distinction" (1961, 139) between thinking of fields as substratum-attribute constructions and thinking of them instead as bundles of property intensities and location properties (or universals, or tropes, or something else). This seems to be asking something like "does the conception of fields we find in contemporary physics more closely fit the philosophically reconstructed ordinary language notion of a substratum with attributes or the philosophically reconstructed ordinary language notion of a bundle of properties," and I do not see how or why the answer to that question adds anything to the conception of such fields that we get from physics itself.

Consider another example:

> An ontological theory of parts and wholes (a mereology) of physical objects describes more fundamental and more general constructional principles than physics or chemistry does, for it gives general principles that govern all the physical objects with parts, including microparticles, atoms, and molecules. For example, chemistry may tell us that the physical structure of a polycarbonate is causally created by arranging elements a certain way, and that its physical parts consist of these arrangements of elements and the attractive forces between them. Mereology contributes the additional claim that the molecule *just is* (say), the mereological fusion of its arranged parts (the elements and the attractive forces). The polycarbonate molecule is created by this mereological fusion, but not in a *causal* sense. Rather, it is created in the compositional or ontological sense: it exists when the parts arranged in the right way exist. So the metaphysics tells us what it *is* to be a sum or physical object composed of these structured arrangements of parts, and thus tells us how the physical object is metaphysically constructed (composed) from its parts. In contrast, chemistry tells us what some of the parts and the arrangements of the parts are for different kinds of molecules, and it also tells us how to causally manipulate the world in order to bring such arrangements into existence. (Paul 2012, 5; original emphasis)

Once we have conceded that chemistry and physics tell us what the parts and arrangements of parts are for different kinds of molecules and how to bring those arrangements in and out of existence, I suggest that there is simply nothing left for a further metaphysical inquiry into the (non-causal) "composition" of molecules or into whether or not they "just are" mereological fusions to be investigating, aside from whether the idea of molecules as such parts in such arrangements strikes us as somehow more intuitively satisfying with or without the conceptual addition of a further relationship of metaphysical or ontological "constitution" among those parts which itself remains obscure (a further puzzle with which metaphysicians will no doubt see it as important to occupy themselves). This seems to be what the appeal to something like inference to the "best" explanation in metaphysical contexts really amounts to, and it is worth noting how closely it returns us to simply reporting what our conceptual and/or linguistic apparatus is presently prepared to count as a "molecule," "physical object," or the like. On this version, we are instead reporting something more like what we would *prefer* to count as a "molecule," "physical object," and so forth, but it still seems profoundly misleading if not perverse to think of this as telling us anything about the actual construction or composition of such entities.

The suspicion that any point of dispute here simply concerns how we prefer to think and talk about the world returns with a vengeance in the context of a further example Paul develops in detail:

> For example, in the debate over whether mereological composition occurs, nihilists argue that for any simples *y*, *y* can be arranged *x*-wise, but there is no sum or composite that is *x*. Thus, there may be simples arranged Finbarr-wise, but there is no sum of these simples that is Finbarr. The view entails that there are no persons, cats, rocks, and stars. Understood in terms of models, nihilists are defending a model of our ordinary beliefs, concepts and language that maximizes simplicity and ontological parsimony at the expense of our commonsense interpretations of what we mean when, e.g., we say that "Finbarr ate the strawberry." The nihilist reinterprets this claim to mean something like *Some simples arranged Finbarr-wise ate the simples arranged strawberry-wise.* Their defense of their model depends on valuing ontological parsimony over commonsense interpretations of ordinary language and other desiderata. Since the nihilist solution also entails solutions to related problems involving composite entities (e.g., there is no problem of material constitution since there are no entities to stand in the material constitution relations) they may also claim that their view has great explanatory value. (2012, 22–23)

Once again, I cannot see a disagreement here as concerned with anything more than which of two ways of thinking and/or talking about the world we find more intuitively pleasing or satisfying. In this example I think distinctively metaphysical inquiry quite evidently consists of simply investigating which further linguistic glosses "we" are most comfortable giving to ordinary linguistic descriptions of the world (like "Finbarr ate the strawberry") rather than settling any substantive matters of fact about the world itself.

A third constituent project of the current metaphysical renaissance, distinct from both scientistic and complementary metaphysics, is something we might instead call the metaphysics *of* science. It seeks to answer questions about the metaphysical commitments *of* the best scientific theories we have: instead of applying science to traditional metaphysical questions, it applies metaphysics to science itself, using the traditional tools and products of metaphysical inquiry to try to better understand what the fundamental assumptions, presuppositions, and commitments of our best scientific theories really are. This is what we do when we carefully scrutinize what our best scientific theories say and how they are put to work to try to decide whether evolutionary theory or quantum mechanics is a fundamentally indeterministic theory, whether or not the Modern Synthesis treats drift and natural selection like forces of the sort we find in Newtonian mechanics, whether the units of inheritance recognized by contemporary genetics are defined by their functions or their material constitutions, and whether and how our best scientific theories make central use of such elements (under various conceptions) as laws of nature, causal dispositions, or natural kinds. Note that these are, at least in the first instance, questions about theories and not about the world: at issue is not whether the *world* is fundamentally indeterministic, for example, but whether the picture of the world given to us *by evolutionary theory* is a fundamentally indeterministic one.[5] To put the difference another way, even if today's scientific theories are ultimately replaced with radically different alternatives even better supported by the evidence, those who work on the metaphysics *of* science will be well-satisfied if they can contribute to this evolutionary process by helping us get a clearer understanding of the scientific theories we already have: what they really say and how they really work. It is simply all the better if we are learning such things about theories that, as it turns out, are not ultimately replaced.

We should note explicitly that there is nothing automatically misconceived or hopeless about this project. History even reveals a healthy number of occasions on which our investigation of some part of the world was indeed productively advanced by explicit reflection on the fundamental metaphysical commitments and implications of our best scientific theories (as when Einstein suggested that the ether was a superfluous posit), and thinkers like Howard Stein (1989), Michael Friedman (2001), and Larry Sklar (2000) have rightly emphasized that much of the thinking of the most creative and theoretically innovative scientists has always been simultaneously scientific and philosophical in character, as those categories are presently understood. But it also seems extremely difficult to tell in advance where, why, when, and how reflection on such metaphysical commitments and implications really will

---

[5] Note that in this way and others the project I describe as "the metaphysics of science" is quite different from that which Mumford and Tugby (2013) have recently associated with the same title. The enterprise they characterize claims to deliver substantive knowledge about the world and seems much closer to what I earlier called "complementary metaphysics": they identify the metaphysics of science as "[t]he metaphysical study of the aspects of reality, such as kindhood, lawhood, causal power, and causation, which impose order on the world and make our scientific disciplines possible . . . and also the study of the metaphysical relationship between the various scientific disciplines" (14).

make some substantive contribution to what I described above as the single, integrated project of naturalistic investigation into the world and our own place within it. We know that often it does not, and this is reflected in the widespread conviction among many working scientists that philosophy is completely and utterly useless in every way. Once in a great while, important progress really is achieved by reflection on the fundamental metaphysical commitments and presuppositions of our best scientific theories, but historically this sort of progress has most frequently been made through such reflection as conducted and guided by the concerns of those who actually do the science. Although philosophers of science are becoming increasingly sensitive to the ways in which making such progress depends on understanding the messy details of the science(s) in question, the pervasive hostility of many or even most scientists toward the contribution of professional metaphysicians to the project of investigating the world itself at least suggests that most of the time the sort of examination undertaken in the metaphysics of science makes little contribution to this integrated naturalistic enterprise, at least of any sort that is broadly recognizable by contemporary scientists themselves.

It may well be, however, that this pervasive hostility ultimately has its source in a further and deeper concern about whether metaphysicians reflecting on the commitments of our best scientific theories can really tell us anything more clearly or more helpfully about those commitments than the original scientific theories themselves do. Indeed, I take some version of this same concern to figure importantly in the motivation for the sorts of broadly "quietist" approaches to science (see Stanford, forthcoming b) offered by thinkers like Arthur Fine (1984, 1986), Stein (1989), and Simon Blackburn (2002). The worry is hard to formulate in a precise and rigorous way, but the central idea is that the best answer to a question like whether evolutionary theory is fundamentally "indeterministic" or treats natural selection as a "force" is *simply to present the claims of the theory* itself. This is arguably more illuminating than trying to decide whether the theory's answer to a question conceived in philosophical terms quite remote from its central concerns is "yes" or "no," or whether the theory even *gives* yes-or-no answers to questions of that sort.

In summary, then, the metaphysics of science is not the sort of fundamentally misguided intellectual project that I have tried to suggest scientistic metaphysics and complementary metaphysics are, but it is nonetheless a project with real pitfalls and challenges of its own to face. Contemporary philosophers of science may indeed ultimately contribute something important to our single, integrated project of naturalistic inquiry by reflecting upon and forcing others to confront the metaphysical commitments and implications of our best scientific theories. But given what seem like very long odds of any particular attempt actually making any such contribution, given the hostility and ridicule routinely heaped by scientists themselves on philosophical inquiry of just this sort, and given the nagging worry that a philosophical restatement of a theory's most fundamental commitments frequently adds little or nothing to the theory's own articulation of those commitments, I think that all we can say to those who are engaged in this project is . . . thank you! Thank you for allowing the rest of us to hedge our bets without doing any of the work, to be confident that with all the time and effort you are putting into this project, if there really is something of value for metaphysical rumination to contribute to the single, integrated, ongoing project of naturalistic inquiry into the world and our own place within it, you will be sure to find it and let us know. If you hear someone in the back of the room shouting "M," "M," "M" after each sentence of your talk, it is probably me, and I apologize in advance. But please understand that I *mean* "M" as an expression of my deepest appreciation, my respect, and my gratitude, for the public service you are performing and for the many sacrifices I suspect you are making in order to perform it.

REFERENCES

Adams, Douglas. 1984. *So Long, and Thanks for All the Fish.* New York: Del Rey (division of Random House).

Barrett, Jeffrey. 2003. "Are Our Best Physical Theories (Probably and/or Approximately) True?." *Philosophy of Science* 70: 1206–18.

Blackburn, Simon. 2002. "Realism: Deconstructing the Debate." *Ratio* 15: 111–33.

Fine, Arthur. 1984. *The Shaky Game: Einstein, Realism, and the Quantum Theory*. Chicago: University of Chicago Press.

Fine, Arthur. 1986. "Unnatural Attitudes: Realist and Instrumentalist Attachments to Science." *Mind* 95: 149–79.

Friedman, Michael. 2001. *Dynamics of Reason*. Chicago: University of Chicago Press.

Hacking, Ian. 1983. *Representing and Intervening: Introductory Topics in the Philosophy of Natural Science.* Cambridge: Cambridge University Press.

Ladyman, James, and Don Ross, with David Spurrett and John Collier. 2007. *Everything Must Go: Metaphysics Naturalized.* Oxford: Oxford University Press.

Lipton, Peter. 1991. *Inference to the Best Explanation*. New York: Routledge.

Mumford, Stephen, and Matthew Tugby, eds. 2013. *Metaphysics and Science*. Oxford: Oxford University Press.

Nagel, Ernst. 1961. *The Structure of Science.* New York: Harcourt, Brace, and World.

Paul, L. A. 2012. "Metaphysics as Modeling: The Handmaiden's Tale." *Philosophical Studies* 160: 1–29.

Quine, W. V. O. 1981. "Five Milestones of Empiricism." In *Theories and Things*. Cambridge, MA: Harvard University Press.

Sklar, Larry. 2000. *Theory and Truth*. New York: Oxford University Press.

Stanford, P. Kyle. Forthcoming a. "Naturalism Without Scientism." In *The Blackwell Companion to Naturalism*, edited by Kelly James Clark, 91–108. New York: Wiley-Blackwell.

Stanford, P. Kyle. Forthcoming b. "Reading Nature: Realist, Instrumentalist, and Quietist Interpretations of Scientific Theories." In *Physical Theory: Method and Interpretation*, edited by Larry Sklar, 94–126. New York: Oxford University Press.
Stein, Howard. 1989. "Yes, but . . . Some Skeptical Remarks on Realism and Anti-Realism." *Dialectica* 43: 47–65.
Worrall, John. 1989. "Structural Realism: The Best of Both Worlds?." *Dialectica* 43: 99–124.

7

An Apology for Naturalized Metaphysics

James Ladyman

1. INTRODUCTION

Analytic metaphysics and science both seek general truths about reality.* Some of the questions they ask about the nature of space and time and matter are at least superficially similar. Given this and the extent of disagreement among metaphysicians about both the answers to their questions and what the questions are, and the high degree of consensus among scientists about both their questions and answers, it is not surprising that some scientists have argued that metaphysics has been replaced by science.[1] Metaphysicians have generally defended their subject by claiming that they address questions that science does not. For example, Jonathan Lowe (2011) and Laurie Paul (2012) have both defended metaphysics against criticisms from other philosophers by arguing that the questions it asks are in some way prior to scientific questions. While Paul talks of fundamental categories and ontological constituents of the world, and Lowe talks of essences, their views about the subject matter of metaphysics and its foundational status are similar. Of course, their accounts raise the question as to how the distinctively metaphysical questions

* The first part of the first chapter of Ladyman and Ross 2007 was written in a polemical and provocative tone and as a result has earned us the opprobrium of many philosophers. Nonetheless, what follows is an apology only in the sense of a defense or vindication. All of what follows is heavily indebted to conversations and correspondence with Don Ross.

[1] Hawking 2010; Krauss 2012.

are to be answered, and Lowe and Paul agree that the method of metaphysics is a priori.[2]

Skeptics about metaphysics, such as the logical positivists (notably Carnap) and many empiricists before (notably Hume) and since (notably van Fraassen 2002), have argued that metaphysics inhibits rather than supports science and is not a worthwhile pursuit. Hume and Carnap regard metaphysical questions as meaningless, or at least fruitless or hopelessly unclear.[3] Van Fraassen accepts that they are meaningful but argues that they cannot be answered either a priori or by science; they all call for an end to metaphysics. In this respect, Ladyman's and Ross's (2007) critique of metaphysics (hereafter "*ETMG*") is very different, since they advocate reform, not abolition.[4] The next section of this paper clarifies and defends our position. First, our negative claims about science and metaphysics are discussed, and some responses to some of them are rebutted. Secondly, some of our most important positive theses are enumerated, and the connection between them is summarized. The related problems of how to characterize both metaphysics and science are explored in the cause of explicating and further motivating the demarcation criterion for naturalized metaphysics we call the "Principle of Naturalistic Closure."

As *ETMG* (chap. 1) discusses, both metaphysicians and scientists are interested in the conditions under which composite objects are composed of simpler ones, yet, astonishingly, while physics and chemistry have made great progress in understanding the composition of matter in both qualitative and quantitative terms, the current vogue in analytic metaphysics is to ignore the relevant science while arguing about whether there are any composite objects at all.[5] Section 3 exemplifies the negative and positive theses of *ETMG* by criticizing the debate about van Inwagen's special composition question, and briefly outlining how composition in condensed matter

---

[2] Paul argues that nonetheless metaphysics uses the same forms of inference as scientists do to choose among empirically equivalent theories; see Ladyman (2012) for a response. Lowe has less to say about how we are supposed to arrive at metaphysical knowledge, but like Sider (2001), thinks that the lack of an adequate epistemology of metaphysics should not stop metaphysicians getting on with their work. Both of them argue that such a lack does not stop mathematicians or scientists, but of course the differences are that, first, mathematicians and scientists agree about the basic truths they have established, and secondly, we do in fact know a lot about how they establish them, whereas in the case of metaphysics there is no common core of metaphysical knowledge nor any reliable methods analogous to proof in mathematics or experiment in science.

[3] Thanks to Galen Strawson for suggesting the latter clause in the light of this passage from and similar in Hume (1740/1975).

[4] They say analytic metaphysics should be discontinued but they have in mind metaphysical inquiry that does not make contact with science, not metaphysics tout court. The term 'analytic metaphysics' names a contingent historical community of philosophers and their writings (see section 2.2).

[5] Korman (2014) explains how eliminativism about ordinary objects is rivalled by the view there are many more objects than common sense or science recognises. (The extensive bibliography makes no mention of the real patterns criterion for distinguishing real objects from spurious composite objects.) See also Wilson 2008 for objections to the metaphysics of composition.

and particle physics is in accord with the real patterns account of existence in which scale relativity and objective modality are central. Section 4 concludes.

2. AN APOLOGY FOR *EVERY THING MUST GO: METAPHYSICS NATURALIZED*

The somewhat infamous critique of metaphysics in *ETMG* differs from the logical positivist critique of metaphysics, both by identifying different problems with metaphysics, and correspondingly by proposing a different remedy. The logical positivists argued that metaphysics is meaningless and should be eliminated. Statements that are empirically unverifiable are metaphysical and have no role in science and philosophy. On the other hand, for Ladyman and Ross metaphysical statements are not meaningless, and metaphysical and scientific thought overlap. However, they argue that the prevailing methodology of current metaphysics, based on applying a kind of cost-benefit analysis to metaphysical hypotheses that are generated without regard to the related science, makes it both unlikely to achieve its own goals, and irrelevant to science, and to those interested in using it to better understand the world. It follows that metaphysics should not be abolished but reformed. In sum, they claim analytic metaphysics is neo-Scholastic, in the pejorative sense of being inward looking, and should be naturalized. That is not to say that they advocate answering all the same questions that are asked by analytic metaphysicians by different means, since they make it clear that they regard some of those questions as meaningful, but as making insufficient contact with reality to be worth entertaining. For example, the question as to under what conditions if any simples, understood as partless particles of matter, compose composites is meaningful, but, since there are no such simples according to our best science, it is pointless for anyone whose goal is to understand reality.[6] On the other hand, the lump/statue debate makes contact with reality in the way ordinary language philosophy does, that is, by addressing the reality of linguistics.[7] It is perfectly possible to generate meaningful discourses about fictional matters. Requiring that metaphysics be meaningful is not a strong constraint.[8]

Even if we restrict ourselves to meaningful questions that everyone agrees are about concrete reality, there may not be any way to answer them. For example, there is no way to know how many hairs Julius Caesar had on his head when he died, and there are many questions about the region of the universe outside our past light

---

[6] The Standard Model of particle physics is a theory of fields that exhibit particle-like behavior, and it cannot be interpreted in terms of anything like classical atomism.

[7] Thanks to Galen Strawson for this example.

[8] Cf. Dan Dennett (2013) on the study of variants of chess that nobody plays and his quip that "if something is not worth doing it is not worth doing well."

cone that we cannot answer because we cannot be in causal contact with the relevant facts.[9] As Alyssa Ney (2014, 133) says, verificationism about truth or meaning may be implausible, but that does not refute verificationism about knowledge, according to which the answers to metaphysical questions are unknowable because empirically unverifiable. What is at issue is the subject matter and methodology, not the meaningfulness of metaphysics.

2.1 *The Negative Content of ETMG*

In respect of the negative content of *ETMG* concerning analytic metaphysics, the main claims include the following:

(1) Analytic metaphysics often involves treating common sense and intuition as sources of evidence about the nature of reality and not merely about the world of appearances and/or our conceptual scheme or schemes.
(2) Science has repeatedly taught us that common sense and intuition are radically mistaken about the nature of reality.
(3) Analytic metaphysics often appears to make contact with science but does so by operating with a "domesticated" version of scientific theories that makes them more amenable to common sense and that interprets them in terms of the manifest image.
(4) Many of the questions that analytic metaphysicians ask are meaningful but are predicated on false presuppositions about the nature of reality derived from domesticated science and/or from common sense and the manifest image.
(5) Collectively, the above premises imply that the methodology of analytic metaphysics does not track the truth and cannot be expected to deliver knowledge of the world.[10]

Cian Dorr's review of *ETMG* (2010) is one of very few sustained attempts to defend analytic metaphysics against our critique.[11] It is endorsed by Brian Weatherson

[9] One does not have to adhere to a causal theory of knowledge to accept that trivial contingent empirical truths cannot be known other than by evidence to which they are causally connected.

[10] Other negative theses in *ETMG* include the claim that analytic metaphysicians often operate with conceptions of causation and individuals that are obviated by our scientific knowledge. Hence the main title of the book that expresses the idea that every 'thing' must go because the world is not made of things in the sense of little material objects as modeled by intuition and as perceived in the manifest image; particles are not intrinsically individuated individuals, nor is there intrinsic individuation by properties (see Ladyman 2015). The nature of particles in physics is discussed in section 3.

[11] See also Katherine Hawley (2010) (Ladyman and Ross reply in the same volume). Heather Dyke and James Maclaurin (2011, 296) reply to Dorr.

and others and seems to have met with general approval.[12] Dorr argues that our criticisms miss the mark because metaphysicians do not after all appeal to common sense or intuitions as a source of evidence or information about how things are. According to his account, we are quite wrong to suppose that common sense or intuition is taken as a reason for belief, rather "saying 'Intuitively, P' is no more than a device for committing oneself to P while signalling that one is not going to provide any further arguments for this claim" (Dorr 2010). Dorr contends that one of the important virtues of analytic metaphysics is the precision with which it investigates claims that have the form of conditionals, and their antecedents need only be assumed for the sake of argument. Hence, he argues against (1) above. According to Dorr, analytic metaphysicians do not generally take intuitions to have probative force. On the contrary, "very [o]ften, 'intuition' talk is playing no such distinctive role," but merely serves to introduce a premise and could be dispensed with in favor of a "pure and chilly" style involving simply their bald assertion.

This is puzzling because there are so many examples in the literature of metaphysicians using talk of what is commonsensical or intuitive as evidence in favor of a particular view. The first chapter of *ETMG* provided various quotations from prominent metaphysicians, and Dorr does not explain why they do not count as evidence for (1). Jonathan Tallant (2014) (who is concerned to defend analytic metaphysics against us) offers various citations from metaphysicians whose work is central to the literature, such as Ned Markosian, Ted Sider, and Dean Zimmerman, explicitly claiming that being intuitive and counterintuitive are epistemically significant features of theories or propositions. For example, Zimmerman (2008, 222) says that being commonsensical counts "very strongly in favour" of a proposition, and Markosian (2008) makes explicit the cost-benefit approach, and among the benefits he adduces, he explicitly includes satisfying intuitions. Indeed, Dorr appeals to intuition in various ways himself in a paper on relations (2004), saying, "I have not been shy of appeals to modal intuitions in this paper: in fact my method of argument relies essentially upon them" (183). Furthermore, at the end of the paper, Dorr offers an explicit statement of the cost-benefit analysis methodology of analytic metaphysics criticized in *ETMG* as follows: "Of course, these advantages must be weighed against the disadvantage of conflicting with our intuition that non-symmetric relations are possible" (191). In the light of all this, it is hard to believe that Dorr and his supporters are ignorant of the role intuitions play in analytic metaphysics; rather it seems that they are in denial about it.

---

[12] http://tar.weatherson.org/2010/06/11/intuition-and-style/ and see http://leiterreports.typepad.com/blog/2012/01/kitcher-on-parfit.html; also Cappelen (2012).

(2) is not at all original to us. *ETMG* cited others making the point most eloquently, and it is not one that has been much contested. However, in another recent paper, Tallant (2013) produces evidence that "intuition" and cognate terms are used a lot in physics. Even if this is true of the sciences more generally, it does not show that scientists use intuition in anything like the way metaphysicians do. Often "intuitive" in mathematics and science is used to mean that something is readily comprehensible using a certain model or set of concepts. Sometimes it is used to describe insight and judgment. In such cases the intuition is schooled—often over very many years—making it a product and repository of existing knowledge. In any case, in science experiment and in mathematics, proof remain the certifiers of community acceptance and theory choice. Intuition is not a determinant of theory choice in anything like the way that it is allegedly deployed in analytic metaphysics, where it is the sole constraint other than logical consistency and theoretical virtues such as simplicity.

If Dorr was right that (1) is false, then (2) would be irrelevant and the argument for (5) in *ETMG* would be seriously undermined. However, of course (5) could still be true and it could be correct that current analytic metaphysics "contributes nothing to human knowledge" (vii). (3) and (4) provide further arguments for (5). *ETMG* offers various examples to substantiate (3) and (4), and in their reviews Kyle Stanford and Paul Humphreys (Stanford et al. 2010) take it to be pretty obvious that we are right about them. Dorr does not explicitly address (5), but he does seem to think that it does not follow from (3) and (4), though he concedes that "there is a fair amount of truth" (1) in our claim that metaphysicians often proceed with false presuppositions. He does not seem to be troubled by this, or to think that it is worth mentioning.[13] This is because he does not regard those presuppositions as really playing any role, since according to him they serve only to provide concreteness and vividness and are dispensable (1). He points out that in the past, scientists made "great contributions to human knowledge despite making many false presuppositions" (1). The difference, of course, is that as science has progressed, agreement has been reached about so much that we can be said to have a large stock of scientific knowledge. The Earth and the other planets orbit the Sun, germs cause disease, the heart pumps the blood to provide oxygen to cells for metabolism, all matter is composed of the elements of the periodic table, and so on. By contrast, as *ETMG* pointed out, metaphysicians continue debates about atomism and universals and cannot agree on any common core of metaphysical truths that have been established. The simplest

[13] Note that Laurie Paul judiciously comments that *ETMG* "raises methodological questions that need to be addressed" (2012, 1), although she disagrees with much of it and is one of those criticized.

way to refute our claim would be give an example of a contribution to human knowledge due to analytic metaphysics, but so far none has been forthcoming.[14]

Of course, the claim that analytic metaphysics contributes nothing to human knowledge is ambiguous, since a contribution to human knowledge could take the form of propositions that can be added to what is known, or it could take the form of indirect contributions such as concepts, heuristics, and models. Indeed, metaphysics has arguably greatly contributed to empirical knowledge, albeit indirectly, for example, in terms of the mechanical philosophy, locality, physicalism, and atomism. Furthermore, some metaphysical systems have arguably been refuted indirectly—Aristotelianism, Cartesianism, vitalism, and, perhaps in the light of Bell's theorem, local realism.[15] This is the account of metaphysics that Peter Godfrey-Smith (2012) offers. According to him, metaphysics is about modeling reality, and so often involves simplifying and idealizing in order to represent the aspects of interest.

Of course, simplification and idealization are often essential to scientific representation, and that per se is not the problem with analytic metaphysics. Rather it is that the models that predominate are based on intuitive pictures of the world based on common sense and the manifest image, and to a lesser extent on domesticated science. This, especially given the current revival of Pre-Socratic, Aristotelian, and Mediaeval metaphysics, makes analytic metaphysics conservative and means that it does not encode much of our scientific knowledge of the world. Hence, it is unlikely to be of any use to individual sciences, which continually seem to require conceptual innovation, as Ladyman and Ross (2007) point out, and it is not helping with the unification of science. The critique of analytic metaphysics in *ETMG* concerns both methodology and content. The method is a priori and out of contact with science, and so accordingly the content tends to involve homely examples of substances and properties and corresponding thought experiments about change of parts and properties over time, or modally.

Much of analytic metaphysics contains no information derived from advanced science, does not engage with the scientific image and questions of interest to scientists, and embodies conceptual conservatism and seeks to domesticate science.

---

[14] In what at the time of writing is the most popular ever post on the blog *MPhi*, Jeffrey Ketland (2012) offers a putative list of such contributions. However, not one of them is due to current analytic metaphysics as targeted by Ladyman and Ross. Bizarrely, Ketland (who does not cite Ladyman and Ross but refers only to recent polemics against analytic metaphysics) responds to an imaginary demand that work in logic, mathematics, and the foundations of science should be discontinued. Both Ross and myself have done much work in the foundations of science and did not call for ourselves to stop doing so. We made it very clear in *ETMG* that it is when metaphysics fails to engage with logic, mathematics, and science that it goes astray.

[15] The "perhaps" is necessary because of the Everett interpretation. For more on the role of metaphysics in science, see Ladyman 2011, 2012.

The analogues of metaphysical hypotheses in science—namely, highly theoretical high-level principles and symmetries—have evolved and are not discussed in the same terms as they were hundreds or thousands of years ago. *ETMG* correspondingly proposes reform of method and content. Of course, there are many metaphysicians and philosophers of science who exemplify naturalistic metaphysics broadly conceived.

2.2 *The Positive Metaphysics of ETMG*

As discussed above, analytic metaphysicians seem largely unmoved by the critique of *ETMG*, and some are happy to argue that metaphysics is a priori first philosophy that underpins science, and that without it there cannot be scientific knowledge at all. On the other hand, some supporters of our negative arguments and views, such as Stanford (2010), object that we do not go far enough by failing to abandon metaphysics altogether. Despite *ETMG* being known to many only as a diatribe about analytic metaphysics, and sometimes being misinterpreted as a repudiation of metaphysics in general, the subtitle of *ETMG* makes it clear that the book advocates and develops a positive metaphysics. It also offers a general conception of naturalized metaphysics that is independent of the specific version the book develops. The critique of analytic metaphysics takes up only a part of one chapter, and the main body of the work is devoted to the construction of a positive metaphysical theory that draws on and attempts to unify our scientific knowledge of the world. The latter is the combination of Ontic Structural Realism due to Ladyman (1998) and Rainforest Realism due to Ross (2000), the avowed aim of which is to offer a theory of ontology to unify physics and the special sciences and to offer an account of causation that is compatible with fundamental physics.[16]

It first defends the idea that insofar as there is metaphysics, it should be naturalized.

(I) Metaphysics should be naturalized and draw upon science.[17]

[16] Dorr (2010) castigates Ladyman and Ross (2007) for lacking the "metaphysician's patience for fine distinctions and quibbling objections" and says that the book fails to make "claims clear and explicit enough to be worthy targets of argument," solely on the basis of having identified several different interpretations of ontic structural realism. It is true that we should have been clearer in this regard, and have been in subsequent work. The claims enumerated in this section are all taken from *ETMG* (albeit unclearly and/or implicitly expressed there). To date they have been engaged with by philosophers of science and scientists more than by metaphysicians. The intellectual commitment to exactness and rigor that Dorr associates with analytic metaphysics are ones that all philosophers and scientists should avow even if they do not always live up to them.

[17] Most philosophers of science have long taken this for granted. Kyle Stanford (this volume) defends what he calls "scientific metaphysics" and made clear in his review of *ETMG* that he agrees with the negative theses of *ETMG*.

In order to give this content, more must be said about what is to count as naturalized metaphysics and to draw upon science. Some philosophers may object that (I) misleadingly implies that metaphysics is distinct from science when in fact they form a continuum and all metaphysical questions are reducible without residue to theoretical questions in particular sciences. On such a view, questions about, say, natural kinds are to be resolved into questions about kinds in biology, chemistry, the social sciences, and so on. One may then wonder what is left that is distinctively metaphysical. Hence, one may endorse (I) while believing that naturalized metaphysics is nothing more than addressing traditional metaphysical questions as they arise in science. Similarly, there are debates about identity and individuality of quantum objects and spacetime points, about biological individuality, about locality, about the nature of the self or free will, that are conducted in the context of specific scientific disciplines. This is clearly a kind of naturalized metaphysics that involves addressing traditional metaphysical questions in a way that is continuous with theoretical debates in science (though one may want to say that this is just science not metaphysics that is just a terminological issue).

The book goes on to argue that

(II) Science is integrated to a very great extent, pace Nancy Cartwright, John Dupre, and others.

Generality is the feature that distinguishes the distinctively metaphysical task from the activity that is continuous with theoretical science. Metaphysics so construed is about the big picture, and the naturalistic version builds this big picture in the light of the actual sciences that we have. What makes the account of naturalized metaphysics in *ETMG* distinctive is the claim that it is about the unification of science and hence it presupposes (II). This motivates a particular account of naturalized metaphysics:

(III) There is a distinctive conception of naturalized metaphysics that involves the unification of science.

The positivists sought to demarcate science from metaphysics by providing criteria of empirical meaning. While some have not given up hope of doing so (Lutz 2012), demarcating science from metaphysics in such a way is fraught with problems. Contemporary philosophers of science tend to have learned from Quine that empirical content is spread throughout our whole conceptual scheme. High-level laws like the conservation of energy have no direct empirical implications without being accompanied by background theories and auxiliary hypotheses, and metaphysical doctrines, for example concerning locality and individuality, have played a role in science. Nonetheless, there is a big difference between most of the hypotheses discussed in current metaphysics and such high-level theoretical principles, since the former are usually completely decoupled from all empirical inquiry and not involved in corresponding scientific debates. Since it is not possible to say a priori what kind of metaphysics is required to unify science, naturalized metaphysics cannot be demarcated by its content or by a prohibition against certain kinds of hypotheses, but should instead be demarcated by whether or not it makes contact with the science of its day and serves the goal of unification expressed in (III) above. Ladyman and Ross express this as follows:

> (IV) Naturalized metaphysics is demarcated by the Principle of Naturalistic Closure (PNC):
>
> If a metaphysical claim is to be taken seriously at a time *t*, it should, if true, show how two or more *scientific hypotheses*, at least one of them *specific* and at least one of them both specific and drawn from fundamental physics taken seriously at *t*, jointly explain more than the sum of what is explained by the two hypotheses taken separately. (Ladyman and Ross 2007, 37)

First, why both the requirements in "at least one of them specific and at least one of them both specific and drawn from fundamental physics taken seriously at t"? If one says "at least one of the Xs is A and at least one of the Xs is both A and B" that is equivalent to saying "at least one of the Xs is both A and B"—the first conjunct is logically redundant. Secondly, why the reference to fundamental physics? There might be things wrong with the metaphysical claim that species are individuals not kinds, but surely it cannot be required to be unifying with respect to fundamental physics? Thirdly, I would welcome a bit of explanation and justification. The text did not really explain what the PNC is for or why we should believe it.

This is intended as a provisional statement and to be understood in conjunction with the account of a very weak notion of the primacy of physics explained in the rest of the chapter. Clearly, it only has content insofar as it draws upon an understanding of "scientific hypotheses." However, the demarcation of science is notoriously problematic, and the history of the debates initiated by the positivists—Popper, Lakatos, Kuhn, and so on—about the nature of the scientific method have not resulted in a consensus about how to characterize science.

Roughly speaking, there are three putative candidates for the nature of science, namely, the content of scientific propositions, the logic of scientific thought, and the social structure of interactions among scientists. It is clear that all these elements are important and that they are not independent of each other at least in practice; however, much philosophy of science has traditionally been focused on the first two. On the other hand, many approaches emphasizing the sociology of science are anti-realist and deny the objectivity of scientific theory choice and progress. Nonetheless, there are forms of social epistemology according to which scientific knowledge is the result of social rationality and is primarily at least possessed by the community, not by individual scientists (for example, Goldman 1999). The collective nature of science is arguably what makes science self-correcting and truth-tracking to the extent that it is, and hence uniquely valued in establishing matters of fact in all other social institutions such as the law, policy making, engineering, and medicine. Ladyman and Ross take it that the only way to demarcate science and metaphysics for the purposes of their polemic is by ostension to the social institutions of science rather than by definition. This is why the PNC is to be understood with reference to "bona fide scientific institutional processes" rather than with respect to a set of necessary or sufficient conditions (Ladyman and Ross 2007, 37).

Metaphysics can similarly be characterized institutionally, and Ladyman and Ross make it clear that their target is analytic metaphysics, and they choose a few authors more or less at random who clearly exemplify that tradition. Their point is not so much that particular individuals are working in violation of the PNC but rather that the center of gravity of the discipline has shifted much too far in an a priori, intuition-led direction that has taken it out of contact with science. Of course, there are many philosophers who practice metaphysics in a way that is informed by science. The problem is that the way that core subjects like the nature of matter, space and time, causation and law, composition and individuality are framed and investigated in analytic metaphysics shows nothing of the scientific image of the world. The PNC is an attempt to say what is required for metaphysics to be informed by science, and for it to contribute to the scientific image.

Many advocates of the unity of science are reductionists, and it might be thought that naturalist metaphysics means materialist or physicalist metaphysics. The question as to how the special sciences relate to physics is of fundamental importance to the positive project of *ETMG*, and again the approach predominately taken to these matters in analytic metaphysics does not make sufficient contact with science. *ETMG* argues that

> (V) There is not for all we know a fundamental level of reality (though there might be), and accordingly our metaphysics should not presuppose that there is.

Accordingly, physics is not defined in terms of fundamental building blocks of reality but as follows:

(VI) Physics is to be characterized as the science that holds at the widest range of scales and across all of space and time.

ETMG argues that

(VII) Reductionism involving either type-type or token-token identity is not in general plausible, though it may be in specific cases.

However, in the absence of reductionism, selective eliminativism is contrary to naturalism, so (assuming some kind of scientific realism which is argued for in chapter 2 of *ETMG*) this motivates Rainforest Realism (following Ross 2000):

(VIII) Science offers us ontologies at many different levels. The ontological commitments of the special sciences should be taken as metaphysically on a par with those of physics. For example, there are atoms, cells, organisms, agents, social structures, and indeed tables.

Furthermore:

(IX) Ontology is scale relative in respect of both space and time.
(X) Composition is a real feature of the world and is in general diachronic, dynamical, and domain specific, since it depends on the relevant kinds of interaction among parts.

These last two claims are discussed in the next section.

ETMG (especially chapters two and three) elaborates a consilience argument for the following claim:

(XI) Structural realism in its ontic not its epistemic form is the most viable position in the scientific realism debate. In particular, neither standard scientific realism nor constructive empiricism is tenable.

This consilience argument makes much of the problems with orthodox metaphysical notions of individuals in physics and other sciences, including biology. (There is of course much more to be said about the characterization of ontic structural realism and the associated debate about individuals in science; see Ladyman 2015, 2016).

It is more or less taken for granted in *ETMG* that

(XII) Scientific representation is often ineliminably mathematical and structural in broad conformity with the semantic approach to scientific theories.

It is argued that rainforest realism and ontic structural realism are unified by the idea that

(XIII) Existence should be explicated in terms of the theory of real patterns.

The definition of real patterns (Ladyman and Ross, 2007, chapter 3) is as follows:

To be is to be a real pattern; and a pattern is real iff it is projectible under at least one physically possible perspective; and it encodes information about at least one structure of events or entities S, where that encoding is more efficient, in information-theoretic terms, than the bit-map encoding of S, and where for at least one of the physically possible perspectives under which the pattern is projectible, there exists an aspect of S that cannot be tracked unless the encoding is recovered from the perspective in question.

The account of real patterns in Dennett, Ross, and *ETMG* is articulated in information-theoretic terms as above. However, Ladyman and Ross (2013) argue that another way to understand it is in terms of statistical structures. Real patterns theory can also be developed in terms of the dynamics of phase spaces (as Jenann Ismael suggested in conversation). Hence:

(XIV) The theory of Real Patterns can be explicated in various ways, including in information-theoretic, statistical, and dynamical terms.

Many real patterns in science are such that the compression of the information allows only probabilistic recovery of the underlying system. Real patterns are lossy, but if one ignores them, one misses out on a real feature of the world. (Consider the Carnot cycle as a real pattern with respect to the underlying statistical mechanics.) Hence, there is not even token-token identity between levels (VII).

Should a naturalistic metaphysician be a realist about natural necessity? Ladyman and Ross argue that objective modal structure is represented by science on the basis of consilience in respect of the following: induction, the no-miracles argument, novel prediction, probability, information processing and computation. Indeed, real patterns are defined modally. They are there to be discovered. Hence:

(XV) The world has an objective modal structure that is represented by causal claims and laws in the special sciences.[18]

Chapter 5 of *ETMG* argues that causation requires a spatiotemporal frame of reference and to that extent at least it is perspectival in something like the sense of Jennan Ismael (this volume, chap. 5). Ismael also explicitly commits herself to there being an objective modal structure to reality. The idea that objective modal structure is a key component of ontic structural realism is emphasized by Ladyman (2000), and reiterated by French and Ladyman (2003). Berenstain and Ladyman (2012) argue that the standard arguments for scientific realism appeal to objective modal structure.

David Wallace (2001) advocates what he calls a "functionalist account of ontology" based on the notion of real patterns in his elucidation of the Everettian interpretation of quantum mechanics. His ontology is two-tier, in that only higher-order entities such as cats and tables are understood in terms of real patterns, whereas the wave function (or whatever else proves to be fundamental in physics) is understood in categorical rather than functional terms. On the other hand, since they hold (V), Ladyman and Ross (2007) advocate a real-patterns account of ontology across the board. There is an asymmetric relation among them due to the fact that physics is maximally general (VI). The relation is not the metaphysical composition of composition as explained in the next section. Ladyman and Ross (2007) deny strong forms of physicalism, since they hold that real patterns at different scales are emergent structure not reducible to the sum of the parts, and that the idea of physical building blocks of higher-level entities is not applicable beyond the atomic scale.

3. COMPOSITION IN METAPHYSICS AND IN SCIENCE

The conception of naturalized metaphysics in *ETMG* is based on Wilfred Sellars's question, "How do things hang together?" (where "things" and "hang together" are understood in the broadest sense of the terms) (Sellars 1963, 1). The idea is that the role of the metaphysician is to say how the sciences as a whole hang together (which presupposes [II] above) in accordance with (III). Assuming that reductionism is not viable (VII), and that we should not be eliminativists about the ontology of the special sciences (VIII), the task is to explain the integration of the sciences and

[18] There are other theses in *ETMG* not mentioned here, but these are certainly among the most important ones.

the relationship between them. The theses above are martialed in the attempt to do so executed in *ETMG*. This section addresses the more limited sense in which "things hang together," namely the fact that there are composite objects composed of parts. According to *ETMG*, scientific accounts of composition are unified by the fact that ontology is scale relative (IX), and that science describes interactions at different scales (X) that can be integrated with each other in accordance with (II). The leads to the idea of real patterns (XIII), which brings together two notions of things "hanging together," namely, necessary connection and composition.

Analytic metaphysicians usually approach composition via van Inwagen's (1990) general and special composition questions. The former is, "What is composition?" and he is skeptical about whether it can be answered. Accordingly, metaphysicians have put much more effort into answering the so-called "special composition question," namely, "Under what circumstances do some objects compose a composite object?" The answers that metaphysicians generally consider to this question assume are (a) that it can be answered without reference to scientific accounts of composition; and (b) that it can be answered in synchronic but not diachronic or dynamical terms. (a) and (b) are related because scientific accounts of composition are usually given in diachronic and dynamical terms. In respect of (a), here is Kriegel:

> If it is true, as I think it is, that our pre-philosophical singular intuitions about objecthood exhaust the data of the theory of composition, then being in accord with such intuitions is in some good sense not just a theoretical virtue of an account of composition (as conservatism is)—it is something more like an empirical virtue. (2008, 363)[19]

Such singular intuitions about objecthood involve objects existing at a time and lead naturally to the synchronic conception of composition (b).

The special sciences concern themselves with what metaphysicians call "sortal relative" composition. Katherine Hawley (2006, 483) explains that the composition question in which metaphysicians are interested is, "What composite objects are there?" This is supposed to be independent of the further facts about what kinds those objects belong to, so that scientific accounts of sortal relative composition are irrelevant. However, there is no reason to suppose that we can know what objects there are without science, nor that how they are composed can be understood without reference to our scientific account of the world. Indeed, Ladyman

[19] Note that this is a clear expression of (1) above.

and Ross asked why we should think that there is an intelligible general question as to what it takes for some things to compose a larger thing, as well as the specific questions answered by the special sciences about when things of one kind compose things of another kind. For example, roughly speaking, chemistry says that atoms compose molecules when there are (relatively enduring) chemical bonds between them, and economics says that agents compose a market when there are financial transactions between them. In both cases, the systems in question are considered over time, and composition is a result of interactions between the parts of the whole. Hence, the answers that science gives to the special composition question make reference to causal processes, not to instantaneous facts about, for example, how close things are to each other.[20]

Up to a point we can very successfully understand the world as composed of smaller spatial parts. However, even in these contexts it is often the case that the parts only compose the whole because of their dynamics (X). For example, it is the interactions among dihydrogen oxide molecules that give aggregates of them the properties of water (see *ETMG*, chap. 1). Composition is diachronic because it typically involves temporal limits and/or relative temporal scales. For example, in statistical mechanics, macroscopic measurements must be carried out over time scales much greater than the mean time taken in atomic interactions. Hence, in science the part/whole relation is shorthand for a process. The scale relativity of ontology is with respect to both time and space because composition is diachronic and dynamical, and because at other length scales, the relevant entities do not exist. For example, there are no water molecules at the scale of quark interactions.

Metaphysical discussions of the special composition question are almost always voiced in the idiom of material particles with some kind of lip service paid to current physics with words like "sub-atomic particles" or "electrons and quarks" disguising the fact that the ontology of the Standard Model and the way particles are described in quantum field theory is incompatible with then wondering whether they are the kinds of things that can form composite systems. There is a whole subdiscipline called "condensed matter physics" that describes how gross matter behaves in terms of interactions between atoms, electrons, and fields. In this context (and in quantum

[20] The obvious arbitrariness of criteria such as proximity in determining composition is one motivation for nihilism, on the one hand, and unrestricted composition, on the other. Both those positions clearly conflict with the scientific image of what there is. Instead of seeking a general conception of composition a priori, those in favor of special composition should draw upon scientific accounts of composition, of which more below. What they have in common is that they explain how a real pattern at the composite level emerges from the real patterns at the level of the parts, so in that sense the theory of real patterns can be thought of as the naturalistic answer to both composition questions.

field theory) the "Renormalisation Group" is the formalism that describes the limiting relationships between theories at different scales. Theories that are renormalizable are such that some physical quantities are independent of the exact length scale cut-off that is made justifying the elimination of some degrees of freedom. There is scale relativity of ontology (IX) because the phenomena associated with the intermediate asymptotic regime exist only at the scales associated with the limit. The universalities—the real patterns—simply do not exist at other scales. More generally, scale-relative perspectives are of vital importance to understanding asymptotic analysis.[21]

Jessica Wilson (2010) argues that the idea of emergent degrees of freedom, in the sense of the parameters needed to specify a state on which the laws depend, is central to the scientific account of ontologies in different sciences.[22] There are numerous examples of emergence in the form of descriptions based on coarse graining with respect to the underlying degrees of freedom that allow a reduction in the effective number of degrees of freedom that we need to use to track the collective behavior of the underlying degrees of freedom. The ideal gas laws use only three degrees of freedom to give a pretty good description of the behavior of systems that have of the order of 10^23 degrees of freedom. The reduction in the number of parameters needed effectively to describe systems is exactly what the theory of real patterns (XIII) aims to capture. It provides a criterion of ontological commitment: real patterns are those that indispensibly figure in projectible generalizations that allow us to predict and explain the behavior of the world.[23] The sciences (and common sense) posit such objects, properties, relations, and processes that allow the formulation of such projections. This is why we should say, *pace* mereological nihilists like Trenton Merricks (2003), that ordinary objects such as tables exist, and why we should deny,

---

[21] This is a ubiquitous feature of physical models, e.g., in the application of the Navier-Stokes equation to the viscosity of the ocean, where the relevant scale is assumed to be small relative to the depth of the ocean so that an infinite limit can be taken. For Bob Batterman (2002), while composition may play a role in the relationship between lower- and higher-level theories, the idea of composition giving rise to novel properties with novel causal powers as considered in much of the literature is a red herring. His analysis of emergence is focused on the existence of a singular limit. In Batterman's example of the caustic singularities in wave optics, there is emergence but no part/whole relation.

[22] Michael Strevens (this volume, chap. 2) defends an account of emergence in terms of reduction in the number of effective degrees of freedom that is similar to the theory of real patterns.

[23] The idea of real patterns is introduced by Daniel Dennett with the example of John Conway's "Game of Life," as explained in *ETMG* chapter 4, in which it is also counseled that various features of the example make it misleading if taken as a metaphysical model of the universe. In particular, in the Game of Life there is clearly a privileged fundamental level of description composed of the aggregation of a finite number of little things, and there is no cross-classification with respect to it at higher levels. However, the example makes it vivid how patterns can emerge at higher levels of description.

pace defenders of unrestricted composition, that arbitrary sums of ordinary objects exist. On this view, there are no real things that do not figure in projectible generalizations/causal laws.

Higher-level laws describe causes that are difference making with respect to the coarse-grained ontology (for example, we may say that the kettle would not have boiled if the gas had not been lit). It is a fact about the world (for all we know a contingent one) that it is ordered on many levels (VII). Science includes theories that link the levels (II), but in almost all cases the higher-level description is coarse-grained, approximate, and aggregative with respect to the underlying levels, so reduction is not plausible (VII). Furthermore, we have learned that reality is not scale invariant. The microworld is not just a smaller version of the macroworld in the sense that the laws of nature seem to care about length scales, energy scale, and velocity. Quantum physics and putatively more fundamental theories describe realms to which the metaphysics of everyday things is not applicable.

Fields can exist in states other than those that can be identified with definite numbers of particles. Indeed, Doreen Fraser (2008) argues that, strictly speaking, there are no particles in quantum field theory. On the other hand, David Wallace (2011) argues that particles are emergent entities. On his account, that there are particles of some kind means that there is an effective quantum field theory involving a Lagrangian for interactions written in terms of particle degrees of freedom. So all particles are really like "quasi-particles." The latter have finite lifetimes, and therefore their existence perfectly exemplifies the scale-relativity of ontology. Physicists describe the world in terms of particles when there are effective degrees of freedom that behave like particles at some scale. Particles such as electrons and quarks are elementary in the sense that they are thought not to decay, but they are not elementary in the metaphysical sense of being excitation states of quantum fields. They also have features that make it quite absurd to wonder a priori under what circumstances if any they form composites aside from the entangled states of electrons. The relationship between quarks and hadrons such as protons and neutrons (nucleons) also illustrates the scale relativity of ontology and the real patterns criterion of existence. The puzzle about quarks is that three quarks can compose baryons, and a quark and antiquark pair can compose mesons, but quarks *are never found free*, and this is a lawlike fact. In high-energy collisions of electrons and nucleons, it seems that quarks hardly interact, yet they are tightly bound together in nucleons. It is posited that the coupling of the strong force in the low-energy limit is infinitely strong, and in the high-energy limit is zero. The "bound state" of quarks that we call the proton has a mass of 938.27MeV even though the component quarks have a combined mass of about 15MeV. Hence, most of the mass, that is the "substance" of protons, is due to the energy associated with the gluons that hold the quarks together. The explanations of all of this have to do with how quarks interact and with scale and the energy regime.

4. CONCLUSION

Analytic metaphysicians have yet to respond to the critique of *ETMG* satisfactorily, and its positive theses are worthy of further exploration. In particular, the account of composition, the scale relativity of ontology, and the theory of real patterns are more plausible than the standard approaches to the special composition question, which itself illustrates how metaphysics can become so decoupled from science as to be pointless. Prima facie composition in condensed matter physics and particle physics fits with the overall account of emergence in terms of real patterns in *ETMG*. Particles exist when there are effective particle degrees of freedom at some scales. Particles are definitely not composed by some general part-whole relation, but are the products of dynamics and interaction among the real patterns of more fundamental physics. Prima facie, things hang together because of the modal structure of the world that is described by the symmetries and laws of scientific theories.

REFERENCES

Batterman, Robert W. 2002. *The Devil in the Details: Asymptotic Reasoning in Explanation, Reduction, and Emergence*. New York: Oxford University Press.

Berenstain, Nora, and James Ladyman. 2012. "Ontic Structural Realism and Modality." In *Structural Realism: Structure, Object and Causality*, edited by E. Landry and D. Rickles, 149–68. The Western Ontario Series in Philosophy of Science 77. Dordrecht: Springer.

Cappelen, Herman. 2013. *Philosophy Without Intuitions*. Oxford: Oxford University Press.

Dennett, Daniel. 2013. "Kinds of Things: Towards a Bestiary of the Manifest Image." In *Scientific Metaphysics*, edited by D. Ross, J. Ladyman, and H. Kincaid, 96–107. Oxford: Oxford University Press.

Dorr, Cian. 2004. "Non-Symmetric Relations." In *Oxford Studies in Metaphysics*, edited by Dean Zimmerman, 1:155–92. Oxford: Oxford University Press.

Dorr, Cian. 2010. Review of *Every Thing Must Go: Metaphysics Naturalized*, by James Ladyman and Don Ross, with David Spurrett and John Collier. *Notre Dame Philosophical Reviews*, June 16. http://ndpr.nd.edu/news/24377/?id=19947.

Fraser, Doreen. 2008. "The Fate of 'Particles' in Quantum Field Theories with Interactions." *Studies in History and Philosophy of Modern Physics* 39: 841–59.

French, Steven, and James Ladyman. 2003. "Remodelling Structural Realism: Quantum Physics and the Metaphysics of Structure." *Synthese* 36: 31–66.

Godfrey-Smith, Peter. 2012. "Metaphysics and the Philosophical Imagination." *Philosophical Studies* 160: 97–113.

Goldman, Alvin I. 1999. *Knowledge in a Social World*. Oxford: Oxford University Press.

Hawking, Stephen, with Leonard Mlodinow. 2010. *The Grand Design: New Answers to the Ultimate Questions of Life*. London: Bantam Press.

Hawley, Katherine. 2006. "Principles of Composition and Criteria of Identity." *Australasian Journal of Philosophy* 84: 481–93.

Hume, David. 1740. *A Treatise of Human Nature*, edited by L. A. Selby-Bigge, 2nd ed. revised by P. H. Nidditch. Oxford: Clarendon Press, 1975.

Ketland, Jeffrey. 2012. "Example of Analytic Metaphysics: Russell's Principles of Mathematics." http://m-phi.blogspot.co.uk/2012/07/example-of-analytic-metaphysics.html.

Korman, Daniel Z. 2014. "Ordinary Objects." *The Stanford Encyclopedia of Philosophy* (Spring 2014 edition), edited by Edward N. Zalta, http://plato.stanford.edu/archives/spr2014/entries/ordinary-objects/

Krauss, Lawrence M. 2012. *A Universe from Nothing*. New York: Simon and Schuster.

Kriegel, Uriah. 2008. "Composition as a Secondary Quality." *Pacific Philosophical Quarterly* 89: 359–83.

Ladyman, James. 1998. "What is structural realism?" *Studies in History and Philosophy of Science Part A* 29 (3): 409–24.

Ladyman, James. 2000. "What's Really Wrong with Constructive Empiricism?: van Fraassen and the Metaphysics of Modality." *The British Journal for the Philosophy of Science* 51 (4): 837–56.

Ladyman, James. 2011. "The Scientistic Stance: The Empirical and Materialist Stances Reconciled." *Synthese* 178: 87–98.

Ladyman, James. 2012. "Science, Metaphysics and Method." *Philosophical Studies* 160: 31–51.

Ladyman, James. 2015. "Are There Individuals in Physics and If So What Are They?" In *Individuals Across the Sciences*, edited by Alexandre Guay and Thomas Pradeau, 193–206. Oxford: Oxford University Press.

Ladyman, James. 2016. "The Foundations of Structuralism and the Metaphysics of Relations." In *The Metaphysics of Relations*, edited by Anna Marmadoro and David Yates, 177–97. Oxford: Oxford University Press.

Ladyman, James, and Don Ross, with David Spurrett and John Collier. 2007. *Every Thing Must Go: Metaphysics Naturalized*. Oxford: Oxford University Press.

Ladyman, James, and Ross, Don. 2013. "The World in the Data." In *Scientific Metaphysics*, edited by Don Ross, James Ladyman, and Harold Kincaid, 108–50. Oxford: Oxford University Press.

Lowe, E. J. 2011. "The Rationality of Metaphysics." *Synthese* 178: 99–109.

Lutz, Sebastian. 2012. "Criteria of Empirical Significance: Foundations, Relations, Applications." PhD diss., Utrecht University.

Merricks, Trenton. 2003. *Objects and Persons*. Oxford: Clarendon Press.

Ney, Alyssa. 2014. *Metaphysics: An Introduction*. Abingdon, UK: Routledge.

Paul, L. A. 2012. "Metaphysics as Modeling: The Handmaiden's Tale." *Philosophical Studies* 160 (1): 1–29.

Ross, Don. 2000. "Rainforest Realism: A Dennettian Theory of Existence." In *Dennett's Philosophy: A Comprehensive Assessment*, edited by Don Ross, Andrew Brook, and David L. Thompson, 147–68. Cambridge, MA: MIT Press.

Sellars, Wilfrid. 1963. *Science, Perception and Reality*. London: Routledge & Kegan Paul Ltd.

Stanford, P. Kyle, Paul Humphreys, Katherine Hawley, James Ladyman, and Don Ross. 2010. "Protecting rainforest realism, Symposium," review of *Every Thing Must Go: Metaphysics Naturalized*, by James Ladyman and Don Ross. *Metascience* 19: 161–85.

Tallant, Jonathan. 2013. "Intuitions in Physics." *Synthese* 190: 2959–80.

Tallant, Jonathan. 2014. "Metaphysics, Intuitions and Physics." *Ratio* 28: 286–301.

Van Fraassen, Bas C. 2002. *The Empirical Stance*. New Haven: Yale University Press.

Van Inwagen, Peter. 1990. *Material Beings*. Ithaca, NY: Cornell University Press.

Wallace, David. 2001. "Emergence of Particles from Bosonic Quantum Field Theory." Available online at http://arxiv.org/pdf/quant-ph/0112149v1.

Wallace, David. 2011. "Taking Particle Physics Seriously: A Critique of the Algebraic Approach to Quantum Field Theory." *Studies in History and Philosophy of Science Part B* 42 (2): 116–125.

Wilson, Mark. 2008. "Beware of the Blob: Cautions for Would-Be Metaphysicians." In *Oxford Studies in Metaphysics*, edited by Dean W. Zimmerman, volume 4. 275–320. New York: Oxford University Press.

Wilson, Jessica. 2010. "Non-reductive Physicalism and Degrees of Freedom." *British Journal for Philosophy of Science* 61: 279–311.

Zimmerman, Dean. 2008. "The Privileged Present: Defending an 'A-Theory' of Time." In *Contemporary Debates in Metaphysics*, edited by Theodore Sider, John Hawthorne, and Dean Zimmerman, 211–25. Oxford: Blackwell.

8

Explanation and Explanationism in Science and Metaphysics

Juha Saatsi

1. INTRODUCTION

Contemporary metaphysics is heavily concerned with its methodology and epistemic status. This is for a good reason: in the virtual absence of experiments, predictions, and empirical feedback, it is far from clear how metaphysical theories and views can be rationally justified. These "meta-level" issues are notoriously as impenetrable as they are important, alas. We can make the task more manageable by focusing on the role and status of *explanatory* considerations in metaphysics. Explanatory arguments and inference to the best explanation play a central, perhaps foundational role in metaphysics. The "explanationist" methodology, although far from being universally adhered to, is widespread and commonly adopted in one form or another. While the notion of metaphysical explanation remains murky and nebulous, and while some have expressed qualms about inference to the best explanation altogether, explanationism has been bravely defended by others. I will argue that these vindications of explanationism in metaphysics turn out to be superficial and vexed, *especially in their affiliation to explanationism in science*. Even if we grant the legitimacy to explanationism in science and everyday life, it turns out to be surprisingly difficult to defend explanationism in metaphysics.

The advocates of inference to the best explanation—"the explanationists"—have much to learn from the study of explanation and explanatory inferences in science. Scientists, like metaphysicians, are also in the business of explaining the world, and explanatory considerations arguably also guide scientific inferences (as the advocates of explanationism are quick to point out). Studies of scientific explanation and explanatory inferences in history and philosophy of science offer a grasp on these issues that has a direct bearing on explanationism in naturalistic metaphysics that is "continuous" with scientific theorizing. The study of explanation and explanatory inferences in science reveals serious shortcomings in prevalent attempts to vindicate explanationism in metaphysics by reference to continuity between science and naturalistic metaphysics.

2. EXPLANATIONISM IN METAPHYSICS

Explanations and explanatory inferences abound in metaphysics; sometimes it is hard to see how metaphysics could even be done if not by comparing potential explanations. I will not question the credentials of all such explanatory reasoning tout court, but rather focus on explanationists who emphasize *the continuity between metaphysics and science*. Here are some exemplars of this more specific target, in alphabetical order.

Armstrong (1983, 1997) famously argues for his account of natural laws via inference to the best explanation. Armstrong also explicitly likens his argument to explanatory inferences to theoretical entities in science. Even induction by reference to laws "becomes a particular case of the inference to explanatory ('theoretical') entities," and "the law, a relation between universals, is a theoretical entity, postulation of which explains the observed phenomena and predicts further observations" (1983, 104). Armstrong ultimately says precious little about the nature and justification of this kind of inference in general; he just maintains that the main explanatory advantage of his laws-as-universals view is *unification*, the relevance of which Armstrong supports by reference to the unificationist theories of scientific explanation.

> We avail ourselves in [metaphysics] of whatever apparently good cannons of explanation we possess or can develop. The basic insight here seems to be involved in the concept of a good explanation: that it should genuinely unify, and that it should be genuinely informative. (Armstrong 1983, 105)
>
> The premise required here is that, as argued by Michael Friedman [and Philip Kitcher], in the natural sciences what, before anything else, counts as a good explanation is something that unifies the phenomena. (Armstrong 1997, 235)

We can continue the list of like-minded explanationists with Bigelow, Colyvan, Dorato, Ellis, and so forth.[1] All these prominent philosophers—mostly from

[1] See, e.g., Bigelow 2010; Bigelow and Pargetter 1990; Colyvan 2006; Dorato 2012; and Ellis 2009.

Australasia, as it happens—have advocated a conception of scientific realism according to which inference to the best explanation can support not only realism about electrons and quarks and the like, but also metaphysical views about, for example, mathematical abstracta, possibilia, laws of nature, the nature of properties and probabilities, and mereology. In the case of Ellis (2009), for instance, the explanationist methodology results in "metaphysics of scientific realism," the ontological count of which includes properties, powers, causes, events, propensities, dispositions, and spatiotemporal and numerical relations, all of which are argued for via inference to the best explanation—a methodology unifying metaphysics and science.[2]

Psillos (2005), an explanationist in philosophy of science, sympathizes with Ellis's explanationism by acknowledging that Ellis's project "rests on the only workable criterion of reality . . . : something is real if its positing plays an indispensable role in the explanation of well-founded phenomena" (398). As Psillos notes, there is a close connection between this "explanatory criterion of reality," also found in Sellars ([1963] 1991), and the well-known indispensability argument for mathematical Platonism. The latter, as championed by its leading contemporary advocates, is again directly associated with explanationism in science.[3]

Swoyer (2008) presents an argument for ontological commitment to mathematical and other abstracta that is quite unlike the indispensability argument. But it is also a forthright instance of explanationism, with a clear emphasis on methodological similarity with science.

> The chief philosophical benefit claimed for . . . abstract entities is that they . . . help explain otherwise puzzling phenomena. . . . [IBE] is not some arcane concoction of meta-physicians. We often infer that something exists on the grounds that its existence would explain something that would otherwise be puzzling. . . . Such inferences also seem common in science. (2008, 16)

Swoyer reflects on the methodology of metaphysics, and for him explanationism represents the answer to "how ontology might be possible" (Swoyer 1999). The idea that explanationism and its methodological affinity with science provide an answer

---

[2] According to Ellis, the test of a good metaphysical hypothesis is twofold: "it must be consistent with the known facts, and be part of a unifying account of reality that explains the overall structure of what we are able to observe" (2009, 123).

[3] See, for example, Colyvan (2006, 229): "I will take the indispensability argument to be an argument that puts pressure on the marriage of scientific realism and nominalism. It does this because the style of argument [viz. IBE] is one which scientific realists already endorse."

to methodological scruples about metaphysics has also been recently defended by Paul (2012), among others.[4]

Explanationism is popular in meta-ethics as well. Here an explicit connection to explanationism in science is drawn by Sturgeon (2006), for example.[5] Examples proliferate; clearly explanationism, with a nod to science, flourishes in metaphysics and beyond.[6] Every philosopher above has their critics, of course, but no wholesale assessment has been given of their common methodological denominator: a widespread justificatory appeal to a similarity between metaphysical theorizing and the explanatory practices of science. There are obvious questions to be asked here: What kind of similarity is this? What kind of justification does it support? What are the limits of such justification of metaphysical views?

I will next characterize in more detail two ways in which philosophers have attempted to capitalize on the similarity between *naturalistic* metaphysics and science, setting the scene for a more critical discussion to take place in sections 4 and 5.

3. NATURALISTIC METAPHYSICS AND SCIENCE

Naturalistic metaphysicians theorize about the natural world—a world also studied by science—by conceiving, assessing, and supporting various theories about it. Many of these theories are admittedly quite speculative. I am interested in the brand of naturalistic metaphysics that recommends, as a methodological dictum, turning to science to see how theories are best assessed and supported. Many naturalists have happily followed Quine in accepting the following three broad methodological

4 See Paul (2012, 22): "This is a central part of my thesis: if we accept inference to the best explanation in ordinary reasoning and in scientific theorizing, we should accept it in metaphysical theorizing."

5 See Sturgeon (2006, 243):

> The justification of many of our beliefs, not just perceptual beliefs, derives from their explanatory role. Thus, for example, the justification of many scientific beliefs is said to lie in their contributing to good explanations of observed evidence . . . and beliefs about others psychological states may explain what we observe of their behaviour. . . . [A]lmost everyone agrees that explanatory coherence . . . is an epistemic virtue in a set of beliefs, contributing other things equal to its being well justified. And this is thought to be especially so when there is explanatory integration across significantly different categories: beliefs about the past explaining evidence in the present, or beliefs about unobservables explaining what we observe, for example. Now, if evaluative beliefs play a plausible role in explaining other facts we have reason to believe in, facts that are not themselves evaluative, then they will be candidates for justification of a similar sort, by their integrative explanatory role.

6 In the sphere of epistemology, Pargetter (1984) advocates explanationism as a response to skepticisms about other minds by viewing inferences to other minds as being relevantly similar to arguments to the best explanation in science. McLaughlin (2010) makes comparable use of inference to the best explanation in philosophy of mind, and Biggs (2011) runs a similar line of thought in relation to modal epistemology.

judgments. First, in science we find a broad methodological injunction to let theoretical virtues guide theorizing, at least regarding matters that are unobservable to our "naked senses."[7] Secondly, in science we can witness (a degree of) confirmation holism: a theory can be related to evidence only as a part of a wider web of theoretical beliefs, not in and of itself. And thirdly, science (suitably interpreted) has all the answers to questions of ontology: there is no "first philosophy" concerning questions of existence. These three naturalist tenets lead to two broad strategies to support explanationism in metaphysics.

The first strategy capitalizes on *methodological similarity* between metaphysics and science in the spirit of the first tenet above. Consider how this tenet is exemplified in the apologia for metaphysics by Sider, Hawthorne, and Zimmerman:

> Scientists must regularly choose between many theories that are consistent with the observed data. Their choices are governed by criteria like simplicity, comprehensiveness, and elegance. This is especially true in very theoretical parts of science, for instance theoretical physics. . . . Just like scientists, metaphysicians begin with observations, albeit quite mundane ones: there are objects, these objects have properties, they last over time, and so on. And just like scientists, metaphysicians go on to construct general theories based on these observations, even though the observations do not logically settle which theory is correct. In doing so, metaphysicians use standards for choosing theories that are like the standards used by scientists (simplicity, comprehensiveness, elegance, and so on). (2008, 7)

Sider, Hawthorne, and Zimmerman stress the idea that in metaphysics, just as in physics, theory choice is governed by theoretical virtues. In effect, they hold that metaphysics is but a *degree* apart from the more speculative reaches of empirical science, and regardless of its highly speculative character, the methodological similarity with science nevertheless justifies the rationality and meaningfulness of metaphysics as a theoretical endeavor. This is a very broad idea, of course, and there are many ways to make it more precise. In particular, one can refine it in explanationist terms, leading to a vindication of explanationism in metaphysics. Thus, Swoyer (2008, 15), for example, appeals to inference to the best explanation in framing a methodological similarity between his arguments for Platonism, on the one hand, and science on the other.

> My suggestion is that we should (re)construe arguments for the existence of abstract entities as inferences to the best over-all available ontological

[7] For Quine, the same theoretical virtues are also important for justifying our views concerning common sense objects.

> explanation. . . . Inference to the best explanation plays a central role in daily life and, according to many philosophers, in science. . . . [M]any maintain, without inferences to the best explanation science, and much of ordinary life, would be impossible.

This is a good example of the strategy that aims to justify explanationism in metaphysics by reference to a methodological uniformity (see also Paul 2012).[8]

As an aside, it is worth nothing that a similar line of thought has been popular in the scientific realism debate in philosophy of science. Here naturalism has been taken to imply that an explanationist argument for scientific realism should exemplify the same method as scientists themselves employ: the realist's inference to the best explanation—empirical success is best explained by approximate truth—should be viewed as a further application of *scientific* explanationism.[9] The realist inference thus arguably exemplifies the same method as various explanatory inferences in science—a fact that arguably provides support for scientific realism via a rule-circular justification of the realist's inference (cf. Psillos 1999, chap. 4; 2011).[10] This, too, is an example of the same strategy for justifying explanationism outside science.

Let us now move on to the second strategy that capitalizes on *confirmational holism*. This flows out of the three tenets of naturalism as follows. In answering questions of ontology, naturalists turn to our best scientific theories, and confirmation holism recommends belief in *all* theoretical assumptions that are responsible for the successes—whether predictive, explanatory, or whatever—that provide realism-eliciting evidence for a given theory. Our grasp of theoretical virtues

---

[8] Note that these philosophers are patently not claiming that metaphysical theories are *equally well* supported by evidence as our scientific theories are. Rather, the claim is that the *kind* of justification is arguably similar to theoretical domains of science. Metaphysics is undeniably much more speculative than science, and one should not object to explanationism in metaphysics—as, e.g., Ladyman (2007, 2012) partly does—on the grounds that it does not have a *probative force* on a par with the explanatory considerations arguably operative in science. For this is just to object to metaphysics' speculative nature, not to explanationism in metaphysics per se.

[9] As Putman puts it, scientific realism should be "viewed . . . as part of the only scientific explanation of the success of science" (1975, 73). Similarly, Psillos maintains that naturalistic philosophers of science "should employ no methods other than those used by the scientists themselves" (1999, 78), and Boyd argues that "the epistemology of empirical science is an empirical science" (1989, 13).

[10] Note that here, too, a naturalistic philosopher is not committed to claiming that her philosophical theory (about science) can be supported to the same degree that scientific theories themselves are supported. One should not object to explanationism in this context merely on the grounds that it does not have probative force on a par with the explanatory inferences in science. This mistake is made by Frost-Arnold (2010), who argues that the realist explanation (and the related inference to the best explanation) fails to satisfy scientific demands for a good explanation (and good explanatory inference), and therefore fails the tenets of naturalism. This demand is based on too strict a conception of naturalism. The realist explanation can be purely philosophical in the sense of not enjoying the degree of *evidence* enjoyed by paradigmatically good scientific explanations. The mode of inference can be the same in the two cases, while the overall evidence (or "epistemological standard") is not.

further shapes our understanding of those successes. This gives rise to a Quinean "indispensability argument": we ought to have commitment to whatever indispensably contributes to our best theories being the best. This then leads to a justification of metaphysical views when it is argued that *assuming* realism about paradigmatic scientific posits (electrons, quarks, etc.) we should *also* be committed to paradigmatically philosophical assumptions (e.g., existence of abstracta, the presentist view of time) on the basis of their allegedly similar role in contributing to our best theories.

This broad idea of confirmational holism can be further refined in different ways. In particular, the notion of indispensability at stake is sometimes framed in explanationist terms, leading to a vindication of explanationism with respect to metaphysical claims. For example, the advocates of the so-called explanatory indispensability argument have recently argued that metaphysical posits (e.g., mathematical and other abstracta) can indispensably contribute to some of our best scientific explanations in a way that is ontologically committing (by scientific realist lights).[11] Such arguments aim to establish that paradigmatically metaphysical views (e.g., Platonism) can enjoy a degree of *empirical confirmation* by virtue of the relevant assumptions (e.g., regarding numbers) playing an appropriate explanatory role—a role that is appropriately continuous with those played by paradigmatically scientific assumptions.

The scene is now set for a more critical discussion. I will not be taking issue with naturalism or the broad idea that science and metaphysics can be viewed "of a piece." Rather, through a closer reflection on explanationism in science, I will only take issue with the explanationist spin that has been put on these two strategies for vindicating naturalistic metaphysics. An incentive for this springs from within the naturalistic conception of philosophy. As said, naturalism recommends turning to science to see how theories are best assessed and supported. But exactly *which* scientific disciplines should we turn to in forming a view of the successful scientific methodology and its limits? I think we should follow Quine himself and understand 'science' quite broadly, so that all scientific findings concerning science and scientists should be taken on board, including those from the history of science and the relevant areas of psychology.[12] After all, there is every reason to expect that these areas of enquiry

---

[11] See, for example, Colyvan (2006, 229): "[IBE] is a special case of the indispensability argument. [T]his is a style of argument that the scientific realist accepts. Mathematical entities surely feature prominently in various explanations." See also Baker (2009) and Psillos (2011) for arguments in the same spirit.

[12] See, e.g., Quine (1995, 49): "I use science broadly, including not only the 'hard sciences' but also 'softer sciences,' from psychology and economics through sociology to history."

have potential to inform us of our capacity and reliability in explanatory reasoning. With this in mind, let us now have a closer look at the two strategies for vindicating explanationism, in the light of the study of explanation and explanatory inferences in psychology and history and philosophy of science.

4. ON THE STRATEGY OF METHODOLOGICAL SIMILARITY

I will now critically examine arguments that defend explanationism in metaphysics by reference to methodological similarity with science (e.g., Swoyer 1999, 2008; Paul 2012). I will put aside various differences in details and focus on a common gambit of methodological unity, which can be expressed in general terms as follows.[13]

The gambit begins with the premise that explanationism is truth-conducive in (some relevant area of) science, and hence justified in that context. It is then argued that (a given area of) metaphysics is methodologically continuous with science in the following sense:

MC1 Both metaphysics and science employ inference to the best explanation.
MC2 We have no reason to think that if explanationism is truth-conducive in science, it is not so in metaphysics.
MC3 We have a positive reason to think that if explanationism is truth-conducive in science, it is also so in metaphysics.

On the basis of these assumptions, the gambit concludes that explanationism in (the given area of) metaphysics is truth-conducive, and hence rational and justified also in that context. (Note that MC2 does not entail MC3: even if we cannot see why explanationism would fail to be truth-conducive only in metaphysics, we may not be able to positively argue for its truth-conduciveness either. Both MC2 and MC3 are required to tackle a skeptic who demands a positive argument for the use of IBE in metaphysics.)

As far as I am concerned, the conclusion follows if MC1–MC3 can be established. But the problem is that MC2 and MC3 have not been established, and there is reason to think that they cannot be established.[14] It is fairly obvious that without support for MC2 and MC3, the gambit reduces to a more-or-less trivial recognition that theory choice in different contexts and disciplines can be *described* in similar explanationist terms at some level of abstraction. It is clear that this kind of purely

[13] Typically explanationists only offer vague *tu quoque* references to science, scientific realism, and inference to the best explanation therein. I construe them as *aspiring* to the schema presented here.

[14] There is reason to be skeptical about MC1 as well. Cf., e.g., Ladyman 2012.

descriptive continuity does not in itself carry any justificatory weight. It is comparable to a foolish attempt to justify any old enumerative induction merely on the basis of it being of the same form as some licit enumerative inductions (cf. Norton 2003).

Before we analyze the problems with MC2 and MC3 in detail, it is worth noting an ill-considered worry about the gambit. It has to do with an appropriate reading of 'truth-conducive' and 'justification' in the schema above. One might think that the gambit is problematic if we have reason think that inferences to the best explanation are *much less reliable* in metaphysics than they are in science, or even if we lack a positive reason for thinking that they are equally reliable in both. For example, one might point to differences in the disciplines' track records—physics vs. metaphysics, say—or argue that our best reasons for thinking that scientific IBEs are reliable do not carry over to metaphysics, raising the specter of potential unreliability (cf. Ladyman 2012). And one might think, in particular, that any such worry about the *relative reliability* in different domains is ipso facto a worry about either MC2 or MC3.

Although there is something to this worry—one cannot wholly divorce a method's reliability from its justification—it is quite difficult to square it with the fact that metaphysics is admittedly inherently *speculative* in a way that our best science arguably is not. I take it that theorizing in metaphysics is generally not taken to be progressive in the way science is. For the realist at least, science is systematically latching onto unobservable reality in an ever-better way, and scientific theorizing is guided or constrained by "correspondence principles" that are grounded in the ideal of discernible continuity in theoretical development. Whatever progress metaphysical theorizing makes, it appears to be compatible with the likelihood of much more significant theoretical discontinuities. Accordingly, our degree of confidence is admittedly significantly lower to any particular output of explanatory reasoning in metaphysics. But this need not mean that in metaphysics explanatory loveliness is not in any sense functioning reliably as a guide to inductive likeliness.[15] For instance, it could be that in metaphysics we are much less able to think of a pool of potential alternative explanations from which we choose the "best," so that we often end up debating over the "best of a bad lot."[16] Still, it could be that inference to the best explanation is quite reliable in picking out the right theory *assuming* that it happens to be included in the pool. In this way the method could

[15] See Lipton (1999, 56ff.) for an explication of the explanationist slogan that "explanatory loveliness guides inductive likeliness."

[16] The "bad lot" objection against explanationism is due to van Fraassen (1989), who employed it against the idea that it is rational to believe that the best explanation is more likely to be (approximately) true than not. The objection loses its bite if explanationism is construed more cautiously, pertaining only to a hypothesis's epistemic probability *relative* to its rival hypotheses (cf. Okasha 2000).

be reliable, for example, "in the long run" (assuming that we eventually manage to conceive of the right theories) and also rational, despite not being reliable in the sense of engendering a high degree of confidence in any particular explanatory inference.

Hence, some difference in the method's relative reliability is compatible with the continuity gambit, because such a difference need not affect the method's justification, which in inherently speculative disciplines such as metaphysics could be based on the notion of reliability "in the long run," or reliability conditional on not having a "bad lot," or reliability in mere relative ranking of alternatives. Unfortunately we have little reason to think that explanationism is reliable even in such a qualified way. As I will next explain, there are other differences between science and metaphysics that give us reason to think that the (assumed) justification of explanationism in science does *not* carry over to metaphysics.

4.1 Issues with MC2

The vindications of explanationism in metaphysics offer some limited support for MC2—the claim that we have no reason to think that if explanationism is truth-conducive in science, it is not so in metaphysics. This support springs from the somewhat vague notion that naturalistic metaphysics and theoretical science are *of a piece*: metaphysics, like theoretical science, is just further (albeit more abstract or general) theorizing about the unobservable world. This naturalist notion that science and metaphysics, as intellectual and theoretical endeavors, are at bottom in the same epistemological boat offers support for MC2 insofar as we have reason to think that there are no differences between these endeavors that would be relevant for the respective reliability of inference to the best explanation.

We *can*, however, point to various sui generis features of science and its explanationist method, and to differences between science and metaphysics, to suggest that MC2 is false: we do have reasons to think that explanationism is not suitably truth-conducive in metaphysics, even if it is (in a sense) truth-conducive in science. For example, we can begin by noting that the truth-conduciveness of explanationism in science can be (and arguably has to be) *qualified* in ways that are not applicable to metaphysics, thus spelling trouble to MC2. The history of science strongly indicates that explanatory inferences in science have (at best) only been reliable in quite limited respects: there is a long and familiar tradition in the history and philosophy of science to point the various gross falsehoods of the best science of yesteryear.[17] Past science has seemingly not been reliable in any significant way regarding

[17] See e.g. Laudan 1981; Stanford 2006; Lyons 2006; Vickers 2013.

its *fundamental ontological posits*, for instance.[18] H
challenge this presents to the explanationist? On
altogether as a guide to deep, unobservable featu
response, MC2 is irrelevant (even if true) giver
response is clearly quite uninteresting in the cur
ing to many scientific realists, it furthermore t
water, since there are ways of qualifying the sense ...
science.

The situation faced by the advocates of scientific explanationism is (p...
somewhat paradoxical: again and again in the history of science it is precisely the metaphysical and ontological assumptions concerning the *nature* of gravity and other forces, light, disease, life, genes, and so forth—the assumptions that were underwriting the best scientific understanding of the relevant phenomena—that have subsequently turned out to be false.[19] The evidence for this claim can be found in the extensive literature on scientific realism (from Laudan [1981] onward).[20] Arguably the best realist response is to accept that the explanationist methodology in science is only reliable in rather limited ways: even if IBE cannot reliably function as a guide to the fundamental nature of things (as the history suggests), it is open for the realist to maintain that scientists, with their IBE-laden methods, nevertheless systematically "latch onto" reality with their theories in ways that largely account

[18] Famous, now-rejected ontological posits include gravitational force, caloric, phlogiston, luminiferous ether, electromagnetic ether, circular inertia, miasma, vortices, vital forces, and electron orbits, to name a few. To be fair, no explanationist has alleged that inference to the best explanation is a sure-fire method, even in science, but stressing the method's *defeasibility* is not a satisfactory response to the worry that the method has been demonstrably unreliable with respect to the reality of fundamental scientific posits.

[19] See Doppelt (2005, 2007) and Saatsi (2012) on the internal tensions that this "paradoxical" situation can present the realist with.

[20] Consider the optical ether, for just one paradigmatic example. It was the assumed existence of the ether that gave the best and only unified understanding of light phenomena for much of the nineteenth century. For a nice discussion, see Cordero (2011), who also catalogues famous testimonies of the resulting confidence in the existence of the ether. These include Maxwell:

> Whatever difficulties we may have in forming a consistent idea of the constitution of the ether, there can be no doubt that the interplanetary and interstellar spaces are not empty, but are occupied by a material substance or body, which is certainly the largest, and probably the most uniform body of which we have any knowledge. (*Encyclopedia Britannica*, [1878] 1893, 9th ed., vol. 8, 572)

and Lord Kelvin (previously Thomson):

> You can imagine particles of something, the thing whose motion constitutes light. This thing we call the luminiferous ether. That is the only substance we are confident of in dynamics. One thing we are sure of, and that is the reality and substantiality of the luminiferous ether. (Thomson 1889, 310; quoted in Thompson 1910, 1035)

and instrumental successes of science. So goes a promising real- the historical "pessimistic induction" over now-rejected theoretical xplanatory mechanisms in "fundamental science."[21]

shot is that even in the face of the pessimistic historical record there is om to consider explanationism in science as "truth-conducive," but only in a r qualified sense. In the light of the history of science, it is difficult to support C2 through the notion that metaphysics and science are methodologically of a piece: the idea that IBE is reliable in science in the way that explanationists would like it to be reliable in metaphysics is arguably premised on a naive view of (the history of) science.[22] Arguably any scientific realism that is sophisticated enough to be tenable in the face of the history of science is highly "selective" regarding our epistemic commitments toward the outputs of the explanationist method, which is only reliable in limited respects. It is a matter of debate how to best characterize the realist commitments and the method's curtailed reliability. While the currently favored realist positions vary significantly in detail, they typically incorporate the broad idea that a (radically) false theory can be empirically successful by virtue of *selectively latching onto* some appropriate features of the world. So, even if scientists' explanatory inferences are not tracking reality in relation to fundamental ontology, a selective realist can save the realist day by claiming that such inferences are nevertheless reliable in ways that fuel scientific progress: they tend to latch onto reality with respect to those aspects of reality that generate the progress and empirical success of science as measured, for example, in terms of increasing empirical adequacy and novel predictions. The debate is subtle, and some care is required lest realism collapses into empiricism, but the overall issue with MC2 is clear: the historical track record impels the realist to adopt the thus-qualified notion of reliability, but it is rather questionable whether it makes any sense to apply such qualifications to explanatory inferences in metaphysics (since the relevant notion of empirical progress seems inapplicable to metaphysics).

The historical track record is not the only reason to worry about MC2. There is a challenge to be faced by naturalistically inclined advocates of explanationism in metaphysics and science alike: to account for the truth-conduciveness of inference to the best explanation in a way that coheres sufficiently well with the rest of our naturalistic world view, according to which our capacities and ways of reasoning about the world are products of natural and cultural evolution, and every reliable mechanism of theorizing is reliable for a natural reason amenable to a scientific study. I am *not* suggesting that the onus is on the explanationist to actually provide

[21] See, e.g., Harker (2013); Saatsi (2016b), and references therein.

[22] This point is also made forcefully by Ladyman (2012).

such a naturalistically acceptable account of the workings of explanatory reasoning. But the challenge nevertheless constrains plausible forms of explanationism, because arguably we should at least be able to *conceive of* a potential account that is naturalistically acceptable. The inability to come up with a how-possible account of the truth-conduciveness of explanatory reasoning reflects badly on explanationism, which starts looking suspiciously mysterious. Here we find another difference between science and metaphysics that gives us a reason to worry about explanationism specifically in metaphysics.

To see the difference, consider first the challenge in relation to science. The challenge is to show that it is not inconceivable how we have developed a capacity for explanatory reasoning that tends to latch onto unobservable reality in ways responsible for the empirical success of science. These ways of thus "latching onto reality" come in many forms. In some cases it is easy to conceive of a naturalistically acceptable story, at least in outline. Consider, for example, the explanatory reasoning that leads to knowledge of the sorts of unobservable *causal relations* that underlie sophisticated scientific interventions exemplified by gene manipulation (see, e.g., Woodward 2010). Discovering such causal connections through explanatory reasoning need not be mysterious because the relevant mode of explanatory practice—causal-mechanistic explanation—is also exemplified in everyday cognitive situations that we are naturally equipped to tackle. According to the new "mechanistic paradigm" in the philosophy of science, much of scientific explanation and associated explanatory reasoning, especially in the life sciences, concern such mechanisms that are not too far-removed from commonplace mechanisms such as engines and clockworks (see, e.g., Bechtel and Abrahamsen 2005). Such causal-mechanistic mode of explanation is, of course, inapplicable to philosophical theorizing.

Other modes of scientific explanation and explanatory reasoning call for a different story, but in general there is something specific to science that helps us to conceive how the "reality tracking" practice of explanatory inferences could evolve even in much more abstract contexts, such as theoretical physics or mathematical biology. In particular, we can appeal to the myriad ways in which *experimental feedback* in science can conceivably shape scientists' domain-dependent judgments of explanatory goodness so as to result in a truth-tracking explanationist methodology (e.g., Kuipers 2002). This is compatible with the kind of explanatory pluralism in science that has been recently emphasized by many (e.g., Douglas 2009), as one would indeed expect a degree of divergence in a detailed evolution of explanationism. It is clearly highly questionable whether these conceivable accounts of the (qualified) reliability of explanationism in science can in any way be extended to inferences to the best explanation in metaphysics, where experimental feedback plays no such role in guiding the explanatory practice.

All this points to a challenge to the metaphysical explanationist that is comparable to the one that Field (1989) (following Benacerraf) has famously posed to the mathematical Platonist. All things considered, it appears much more difficult to conceive of a naturalistically acceptable account of the truth-conduciveness of explanationism in metaphysics. The onus is on the explanationist to show that we can conceive of an account of our reliability in explanatory inferences concerning, say, matters of fundamental ontology or modality. Broad references to explanationism in science pay mere lip service to this challenge given the disanalogies noted above.

As an aside, it is worth noting that one of the most prominent scientific realist arguments is also equally targeted by the above criticism. The global realist meta-abduction argument ("no-miracles argument"), as developed by Psillos (1999) and others, is a profoundly problematic exercise in philosophical explanationism. At stake is a "global" reliability of the abductive method of science, which is supported via an inference to the best explanation, a philosophical inference that is continuous with explanationism in science.[23] In my view, the global no-miracles argument fails to bridge the gap between scientific and philosophical explanationism. The fact that the explanandum at stake is *about* science does nothing to diminish the above worries about the continuity gambit exhibited also by this realist argument.[24]

So far I have surveyed reasons for thinking that explanationism, even if truth-conducive (in a qualified sense) in science, is not so in metaphysics. Next I will turn a critical eye to the positive arguments for thinking that the presumed truth-conduciveness of explanationism in science carries over to metaphysics.

4.2 Issues with MC3

What works for the goose, works for the gander. Or so the advocates of explanationism think: if explanationism is truth-conducive in science, it is in metaphysics also. Positive arguments to this effect typically point to underlying similarities between scientific and metaphysical explanations and the respective explanatory virtues. But what little has been said of the pertinent similarities turns out to be problematic in the light of current analyses of scientific explanation (insofar as those analyses are not applicable to explanations in metaphysics, in particular).

---

[23] Cf. section 3 and note 11.

[24] Since I am not concerned with defending scientific realism here, I refrain from exploring better vindications of scientific explanationism. See Saatsi (2009) for further discussion and a recommendation that a realist justification of scientific explanationism should proceed "piecemeal" (as opposed to "wholesale," as in the case of no-miracles argument), relying more heavily on the role of experimental feedback that shapes our explanatory judgments.

Swoyer (2008) typifies the slender attempts to support MC3. After noting an obvious disanalogy between scientific and philosophical explanations—only the former typically involve causal mechanisms—Swoyer points to the relevant analogies:

> There are a number of [common] explanatory virtues in addition to pinpointing causal mechanisms, however, including unification, integration, and redescribing phenomena in a theoretically enlightening way. Seeing a common pattern at work behind seemingly disparate phenomena certainly yields one sort of understanding. By describing the moon, the earth, and projectiles on the surface of the earth as bodies with inertial and gravitational mass, Newton was able to provide a unified account of their motions. Something similar can occur in philosophy. (27)

Swoyer is not alone in pointing to *unification* and explanations' *power to give understanding* as relevant features of good explanations, shared by metaphysical and scientific explanations alike. Typically these explanatory virtues are characterized in extremely general and abstract terms. But on closer inspection, we will see that broad references to these features pay a mere lip service to the literature on scientific explanation, which provides plenty of reason to think that no substantive overarching account of such measures of explanatory goodness can be given to unify metaphysical and scientific theorizing as desired.

Understanding. Let us focus on explanations' "power to give understanding" to begin with. (I will discuss unification separately below.) Is it safe to assume that theories' power to give understanding is a truth-conducive epistemic virtue that can guide theory choice, relating epistemic agents to the unobservable reality regardless of their domain of enquiry, so as to support MC3?[25] Such an assumption is often tacitly made, without evidence, by the explanationists.

We should worry about the validity of this tacit assumption. Without clear understanding of "understanding" (especially in philosophy), one may be inclined to regard "power to yield understanding" as too subjective and psychological a notion to support a wholesale justification of explanationism (unspecific to a domain of enquiry). This initial worry is strengthened by the fact that science itself turns out to exhibit a huge variation in its theories' claimed "intelligibility" across different historical and social contexts and across different domains

---

[25] By "domain of enquiry" I denote not only the subject matter of theoretical enquiry, but rather the subject matter in combination with the myriad psychological and social factors that shape the study of that subject matter. Thus, ancient Chinese astronomy constitutes a different domain of enquiry from twentieth-century Western astronomy, for example.

of theorizing (e.g., Kuhn 1977; de Regt and Dieks 2005). Our knowledge of the scientific endeavor indicates that scientists' evaluation of explanations in terms of their provision of understanding can be heavily influenced by various kinds of social and cultural factors; it is not simply a matter of determining some objective feature of explanations.

For example, some excellent scientific theories in the history of science have lacked the power to provide understanding to leading contemporary scientists. For a well-known example, Huygens, Leibniz, and much of the scientific community regarded Newton's universal gravitation as involving an unintelligible action at a distance—a sentiment echoed by Newton himself (at times). By comparison, an average physics student today regards Newtonian gravity as a perfectly cogent explanatory posit. Sometimes the intelligibility of a new theory eludes everyone apart from its architects, as when the physicists in the Copenhagen-Göttingen circle around 1926 claimed to understand quantum phenomena with matrix mechanics, while most other physicists outside the circle regarded it unintelligible (de Regt 2001). There are also clear disagreements across different scientific fields regarding the level of understanding provided by different theories and different theoretical approaches. For instance, it is not uncommon for a group of mathematical biologists to claim that an abstract mathematical theory (e.g., of biological scaling laws) gives them deep understanding of biological phenomena, only to be faced by severe dissent from other camps of biology (see, e.g., Keller 2002).

This kind of demonstrable variability in scientific explanations' intelligibility and power to give understanding speaks against the tacit assumption made in support of MC3. In science it can be impossible to objectively rank theories with respect to the level of understanding they provide, given the way in which "understanding" is interwoven with theorists' social, cultural, and philosophical background.[26] Why would "understanding" be any more objective affair in metaphysics?

To be clear, those who defend explanationism in science must of course also accommodate these data. But the realist arguments to this end can at least appeal to specific features of science, such as experimental feedback and an objective sense

[26] Such objective ranking eludes us, at least insofar as "the level of understanding" provided is meant to be metacognitively accessible to theorists in question. In the context of specific accounts of scientific explanation, we can try to put "understanding" on a more objective footing by construing "genuine understanding" (as opposed to mere "sense of understanding") as grasping of a genuine explanation. Such "genuine understanding" may bear no relation to agents' own assessment of their understanding, however. If an account of explanation offers a way to measure "explanatory depth," it can be furthermore associated with the level of genuine understanding. For example, in relation to Woodward's account of scientific explanation, understanding has been construed by some as the ability to draw correct counterfactual inferences, rendering "the level of understanding" proportional the measure of inferences one can make (Ylikoski and Kuorikoski 2010). It is difficult to apply this to philosophical explanations, however.

of scientific progress. Such arguments, being specific to science, are of no help to an advocate of MC3, however.

The historical and social studies of science cannot be ignored by a naturalistic philosopher who (in the Quinean spirit) views philosophy as continuous and in harmony with our *overarching* vision of science and scientists. Psychology and cognitive science form an ineliminable part of that vision as well. Some philosophers of science have recently turned to these disciplines to explore the relevance of *scientific* studies of explanation and understanding, resulting in increasingly naturalistic perspectives on these topics, yielding hitherto unexplored challenges to explanationism in metaphysics.

Studies of the psychological state of understandings have revealed serious issues with invoking the *sense of understanding* offered by an explanation as a reason for accepting it. In his review of these issues, Trout (2007, 566) argues that "the research findings on judgment and decision-making [indicate that] the sense of understanding is a common, but routinely unreliable, index of intellectual achievement." According to Trout, this is not just a matter of reduced reliability; rather, "there is little practical wisdom in relying on the sense of understanding as a cue of a good explanation, or even a potential one" (574). Trout argues for this by pointing to two well-documented psychological biases (hindsight and overconfidence), as well as other robust psychological phenomena.[27]

Trout primarily aims his challenge at scientists and scientific realists, but it is obvious that the challenge applies to explanationism at large.[28] If the sense of understanding evoked by an explanation is influenced by psychological biases, and we are operating in a domain that does not return any systematic feedback on our explanatory judgments, we ought to worry about the grounds of basing an inference on understanding. Furthermore, as Trout goes on to discuss on a more positive note, there *are* some responses available to a scientific realist (qua explanationist), hinging on the role of explanatory considerations in the empirical progress of science.[29] But these responses are difficult to extend to explanationism in metaphysics, where

[27] See also Grimm (2009) for a more optimistic vision on the reliability of understanding.

[28] In some domains, scientists have grown acutely aware of these issues (e.g., the problem of just-so stories in evolutionary biology), in other domains less so (e.g., the problem with the so-called rational addiction theories in economics; cf. Rogeberg 2004).

[29] See Trout (2007, 586):

> Scientific realists can assign a robust role to objective factors in explanation—such as statistical and causal relevance—and value the contribution of explanation to scientific progress. If your focus is not balanced by a positive account of the sense of understanding in a scientific theory of explanation, then it is easy to portray any criticism of the sense of understanding as the first step toward explanatory nihilism. But explanatory nihilism is surely premature.

the critical notion of empirical progress—involving predictive and instrumental successes—is simply inapplicable.

In a spirit similar to that of Trout, Lombrozo (2011) reviews the growing literature within psychology and cognitive science that attests to the *instrumental value* of explanations for everyday cognition, with lessons that arguably generalize to science. According to the research reviewed by Lombrozo, explanation can have instrumental value in everyday cognitive tasks of generating predictively useful theories and promoting the discovery of causal structures, for example, by virtue of its intimate cognitive relationship to learning and causal inferences. According to the picture painted by this research, the cognitive mechanisms shaped for *our* everyday "theorizing" and learning naturally carry over to *our* practice of science (where these mechanisms get refined by empirical feedback—e.g., confirmed predictions—that can contextually shape our grasp of explanatory virtues, for example). These cognitive mechanisms also naturally get applied to metaphysics, no doubt. But while this line of thought can to some extent rationalize the explanatory endeavors of science, the talk of the "instrumental value of explanation" in metaphysics is tenuous at best, and in the absence of empirical feedback to guide our explanatory practices, it is much more difficult to justify the truth-conduciveness of those practices, especially when they are far-removed from their cognitive origins. (This issue connects to the challenge discussed toward the end of section 4.1.)

Unification. Unification is one of the most often cited (alleged) good-making features of both scientific and philosophical explanations.[30] The advocates of explanationism also often regard unification (ceteris paribus) as a universally recognizable objective explanatory virtue of a theory, or even as a constitutive feature of explanation, as Friedman and Kitcher have suggested.[31] Thus, the thought goes, we can arguably assess philosophical theories with some reliability by virtue of being able to gauge relative degrees of unification.

However, in light of contemporary philosophy of science, the relationship between unification and explanation is much more delicate in ways that threaten to undermine broad appeal to unification in support of MC3. First of all, the attempts (by Friedman, Kitcher) to view unification as a *constitutive* feature of explanation have faced severe criticisms and are highly contentious.[32] And although there is an important seed of truth in the weaker idea that unification is an explanatory *virtue*,

---

30 See Swoyer (1999, 2008); Armstrong (1997, 235–36); and Sider (2007), for example. Some regard the argument for genuine modal realism in Lewis (1986) also critically hinging on unification in a similar spirit. Ditto Sider (2011).

31 Both Friedman and Kitcher also associate unification closely with understanding. Cf. Friedman (1974, 15) and Kitcher (1989, 432).

32 See, e.g., Gijsbers 2007; Barnes 1992; and Woodward 2003, chap. 8.

the way this idea is developed in different contemporary accounts of explanation offers no solace for MC3.

For example, a well-received, recent counterfactual theory of explanation offers a promising way to construe unification as an explanatory virtue. In this account, unification comes out, first of all, as one of *several competing* virtues, and secondly, as a good-making feature of *individual explanations*, not of scientific *theories* per se. Let me elaborate. We can begin, with Woodward and Hitchcock (2003), by recognizing that explanatory power is naturally associated with the *generality* of an explanatory regularity employed. The basic intuition harkens back to the covering-law model of explanation, according to which an explanandum is explained by subsuming it under law-like explanatory regularities. The more these laws cover—the more general they are—the deeper the explanation, according to the intuition. Unificationist accounts of explanation took this idea as their starting point, but arguably misidentified the sense of generality at stake. As Woodward and Hitchcock put it,

> [T]raditional approaches to explanation have ... focused on the wrong sort of generality: generality with respect to objects or systems other than the one that is the focus of explanation. The right sort of generality is rather generality with respect to other possible properties of the very object or system that is the focus of explanation. (2003, 182)

In their account, explanation is a matter of exhibiting a pattern of counterfactual dependence that describes how the explanandum would change under various conditions. Explanations provide such modal information in terms of generalizations that are *invariant* under appropriate changes in some variable values (indexing determinate properties).[33] It is a major attraction of the account that it provides the means to capture explanatory depth in general terms, in terms of the *range of invariance* of a generalization (cf. Woodward and Hitchcock 2003). Unification, furthermore, comes out as an explanatory virtue to the extent it can be linked to generalizations' degrees of invariance. Far from being a *distinctive* explanatory virtue, however, unification comes out as one of at least seven different "dimensions" of explanatory depth, namely, different ways in which generalizations' degree of invariance can vary (Woodward and Hitchcock 2003). It can be furthermore argued that some of these different dimensions are *systematically in conflict*, and that scientists' weighing of different dimensions—their assessment of explanatory goodness, in

[33] For example, Newton's gravitational law is invariant under changes in the specific masses and their relative distance.

other words—is typically contextual and pragmatic in nature (Ylikoski and Kuorikoski 2010).

Arguably, we can account in these terms for the prevailing general intuition (supporting MC3) that unification is an explanatory virtue (Ylikoski and Kuorikoski 2010). We can also specify more precisely the conditions under which the intuition holds.[34] The upshot is that unification is connected to explanatory goodness in a rather specific and less objective way than assumed by the advocates of MC3. Furthermore, the notion that we can compare different *theories* in relation to this explanatory virtue sits somewhat uncomfortably in the Woodwardian framework, which is primarily concerned with analyzing the depth of *individual explanations*. Thus, we can make good sense of the notion that an explanation of Uranus's orbit in terms of general relativity is better than in terms of Newtonian mechanics. Or that the latter explanation does not improve if it cites the ideal gas law in conjunction with Newton's law, even though it (artificially) "unifies" two types of phenomena in the covering-law sense. But for this we do not need the notion that Einstein's *theory* is more explanatory than Newton's by virtue of being more unifying. The notion that general relativity as a theory has more "explanatory power" than Newtonian mechanics may or may not make derivative sense on the basis general relativity in general furnishing explanations that are individually better in the relevant sense.[35] But such derivative (and seemingly unneeded) notion of explanatory power is intelligible (at best) when the individual explanations can in the first place be compared in the Woodwardian sense. This particular analysis of explanatory unification thus bestows no general notion of unification-qua-explanatory-virtue applicable outside the province of Woodward's account. Thus, insofar as Woodward's account captures the unificatory virtue of scientific explanations, and insofar as this account fails to capture the sense in which metaphysical theories explain, one does not find in science support for the dictum that in general the most unified theories—whether scientific or metaphysical—are the likeliest to be true since they (ceteris paribus) explain the best.

---

[34] The explanatory power of an explanation is increased by integrating it into a unifying theory *only if* it increases the amount of explanatory modal information provided. According to Ylikoski and Kuorikoski (2010, 214),

> [Such] integration [should] be more than formal compatibility or conceptual coherence. The bodies of knowledge should come together in a manner that allows new relevant inferences to be made about the phenomenon. The requirement of relevance makes the idea of integration a local notion: it depends on one's epistemic aims. This contextual notion is quite different from the global notion of unification advocated by the supporters of unification accounts of explanation.

Furthermore, arguably such integration can often be achieved only by compromising other dimensions of explanatory depth, for example by idealizing away fine-grained detail and accuracy.

[35] Cf. Ylikoski and Kuorikoski 2010.

I have flagged Woodward's account as an important example of contemporary philosophy of science that sheds light on scientific explanation in a way that considerably complicates the simple picture underlying MC3, according to which scientists appeal to unification as a theoretical virtue much like metaphysicians do, via inference to the best explanation. Woodward's account is not the only game in town, of course, but other notable contemporary accounts of explanation are equally hostile to the idea that unification is an explanatory virtue of a *theory*. The kairetic account of Strevens (2008), for example, values generality in some very specific ways in locating its explanatory difference-makers, but it does not associate any kind of explanatory virtue with unification. So also in light of this recent account of scientific explanation, there is reason to worry about the commonplace idea that metaphysical theories can be appraised in terms of their unifying explanatory power simply "because scientific explanations are so appraised."

Of course, none of the above discussion goes to show that unification is not a *theoretical* virtue in science. But even if it is—even if more unified theories are generally better theories, ceteris paribus, by the lights of science—the question is whether unification should be thought of as an *epistemic* or merely *pragmatic* virtue. Here again it is natural to turn to the history of science, which indicates that unification is often best thought of as a pragmatic virtue, not as an epistemic guide to what the world is like. For example, in her book-length study of unification, explanation, and realism in science, Morrison (2000) points to various theoretical unifications in physics and biology that are not explanatory (as she argues), or at least were achieved by means of mathematical structures and theoretical constructs that do not reflect reality.[36] The truth is that the historical record attested by Morrison is a mixed bag: sometimes theoretically virtuous unification is a guide to ontology (e.g., Maxwell on electricity and magnetism), but often it is not (e.g., the electroweak theory, or Kaluza-Klein theory).[37] Given this lesson from the history, it is only right that in connection with some exemplar cases scientists themselves have not been epistemically swayed by mere unification, but have demanded *independent evidence*, as in the case of Maxwell's theory, the vindication of which had more to do with

---

[36] Morrison (2000, 232) sums up her take-home message thus:

> [N]o single account of theory unification can be given. A philosophical consequence of that claim is that unity should not be linked to truth or increased likelihood of truth; unification cannot function as an inference ticket.
>
> There can be no doubt that unity exists in science, that unified theories have been enormously successful and that unity is a goal pursued by many practicing scientists in a variety of fields. But nothing about a corresponding unity in nature follows from those facts.

[37] See also Karaca 2012; Maudlin 1996.

the independent evidence for electromagnetic waves from Hertz's experiments conducted some years after Maxwell's unification.

Contemporary philosophy of scientific explanation paints a fascinating naturalistic picture, the details of which challenge in various ways the *strategy of methodological similarity*. This strategy, recall, aims to justify explanationism in metaphysics on the basis of its role and (assumed) epistemic status in science. For the sake of the argument, I have granted that explanationism in science is rational and reliable, although obviously the sense of reliability has to be qualified in quite specific ways to fit the picture painted above. Can scientific realists actually defend a sufficiently robust, yet suitably qualified reliability of explanationism in science? Whatever the answer, we have reviewed a number of relevant aspects of explanation and explanationism in science that collectively throw considerable doubt on the abstract vindications of metaphysical explanationism.

The explanationist can retreat: instead of trying to justify *metaphysical* instances of inference to the best explanation, she can argue that some metaphysical views get support directly through *scientific* instances of inference to the best explanation. This is the gambit of confirmational holism to which I now turn.

5. ON THE STRATEGY OF CONFIRMATIONAL HOLISM

Naturalistic metaphysics sometimes engages in exploring the ontological commitments of science, as in the case of the so-called indispensability arguments. I will now examine the connection between these arguments and scientific explanationism.

The best-known indispensability argument views mathematical realism as a *corollary* of scientific realism (Colyvan 2006). Briefly and roughly put, in the Quinean holistic picture of science, the empirical justification for theoretical assumptions "bleeds over" to (applied) mathematics. As Psillos (2012, 53) puts it,

> [I]ndispensability arguments capitalise on the strengths of scientific realism, and in particular of the no-miracles argument (NMA), in order to suggest that a) the reality of mathematical entities (in their full abstractness) follows from the truth of (literally understood) scientific theories; and b) there are good reasons to take certain theories to be true.

An explanationist can also add that empirical justification is "mediated" via inference to the best explanation, as this is how we can arguably construe the relation between our best theories and their empirical support. Furthermore, the argument for scientific realism can itself be an instance of inference to the best explanation, as in the case of the no-miracles (NMA) argument:

> The epistemic optimism characteristic of scientific realism is based on NMA. The argument, roughly put, is that empirical success (suitably regimented so as to include novel predictions and the like) offers good reasons to believe in the truth of theories, since it is best explained by the claim that theories are true. Thus conceived, NMA is blind to a distinction between abstract entities and concrete ones insofar as commitment to both types is implied by the truth of (literally understood) scientific theories. (Psillos 2012, 53)

The original (Quine-Putnam) indispensability argument, recited by Psillos, concerns ontological commitment of *literally true* theories. What should we make of the fact that in light of the above discussion we are not justified in taking explanatory inferences in science as delivering literal truths? If mathematical realism is but a corollary of scientific realism, presumably it matters what kind of scientific realism it is rational for us to maintain? Taking on board our complete understanding of science and its explanatory endeavors, we must admit that the extent of our ability to answer ontological questions with science, naturalistically, depends on whether or not, and in what sense, scientific explanationism is reliable. The history of science indicates that the reliability of explanationism is curtailed: explanationism allows us to track the unobservable reality only in limited respects (cf. section 4.1). Such curtailed reliability can still generate considerable empirical successes (including novel predictions and the like), if the outputs of explanatory inferences selectively latch onto appropriate features of the world—namely those features that entail the right predictions, and so forth.[38] Hence, the sort of no-miracles argument alluded to by Psillos above is *over*-optimistic, since the best realist explanations of the empirical success of science need not be in terms of the *literal* truth of theories. In the history of science, the realist must repeatedly explain the empirical success of theories that are only partially veridical, in a way that is compatible with the curtailed reliability of the scientific method.

We have reason to worry about indispensability arguments, since we have reason to worry that realist explanations of empirical success fail to support mathematical realism.[39] It is a considerable, hitherto unsettled question whether there is any

---

[38] For example, although Fresnel's ether theories of optics or Newtonian gravity are far from being literally true, and their explanatory posits do not exist, these theories arguably selectively latch onto reality so as to generate hugely impressive empirical successes.

[39] I think Hawley is methodologically on the right track in thinking that in naturalistic metaphysics "we have reason to believe metaphysical claims to the extent that they are *genuinely involved* in generating empirical success"

support for mathematical realism (or other paradigmatically metaphysical views, such as presentism) to be found in the best scientific realist arguments. I am skeptical, but I will not argue for this here. My present point is that our overarching understanding of scientific explanation and explanationism can challenge the strategy of confirmational holism according to which the scientific realist should by her own lights be committed to paradigmatically metaphysical views.

The recent literature on the indispensability argument displays a broad consensus according to which scientific realist commitments cannot be read off from our best theories construed as true *simpliciter*. Philosophers are nowadays much less concerned with mathematics' *unqualified* indispensability in science—the unvarnished Quinean notion that our best scientific theories simply quantify over mathematics. The focus has rather shifted to mathematics' indispensability for scientific *explanations*.[40] This shift presents a further explanationist twist on the strategy of confirmational holism.[41] The champions of the new "explanatory indispensability argument" envisage a more direct route from scientific realism to mathematical realism, based on the notion that (for a scientific realist) ontological commitment and scientific explanation are directly connected. Realist commitment to mathematics allegedly follows from the admission that mathematics plays a genuine "explanatory role" in science.

Although the focus on explanation admittedly enhances the indispensability argument, our best understanding of scientific explanation yet again considerably complicates the key issue at stake. The (burgeoning) literature on the explanatory indispensability argument has by and large taken for granted the connection between explanatory indispensability and ontological commitment, without any reference to a particular conception of explanation to underwrite this connection. Prima facie innocent assumptions concerning explanations' ontological commitments turn out to harbor various complications, however, when examined in relation to specific accounts of explanation. In particular, the literature on the

(2006, 460; my emphasis). Hawley argues that the indispensable role of metaphysical assumptions in science can be revealed by their involvement in the best realist explanations of some empirical successes of science. But while I broadly agree with Hawley's methodological pronouncement, I see no grounds for her optimistic belief that "there are actual cases in which the involvement of a metaphysical claim in an empirically successful scientific theory provides some reason to think that the claim is true" (456).

[40] See, e.g., Colyvan 2012.

[41] Baker (2005, 225) puts it thus:

> A crucial plank of the scientific realist position involves inference to the best explanation (IBE) to justify the postulation in particular cases of unobservable theoretical entities. . . . [T]he indispensability debate only gets off the ground if both sides take IBE seriously, which suggests that explanation is of key importance in this debate.

indispensability argument is rife with references to "mathematics' (indispensable) explanatory role," but the key notion of "explanatory role" has been mostly left unanalyzed. This is a major shortcoming: the notion of "mathematics' explanatory role" must be examined in relation to different analyses and conceptions of explanation to properly judge whether mathematics plays *the kind of* ontologically committing explanatory role that matters for the indispensability argument. This is critical because in the context of different accounts of explanation we can delineate *different kinds* of explanatory roles. The admission that mathematics is in *a* sense "genuinely explanatory," or is "playing an explanatory role," does not by itself at all imply that it is playing the *right* kind of explanatory role. Saatsi (2016a) draws some critical distinctions between different types of explanatory roles—in connection with some leading "ontic" accounts of explanation in particular—pointing to various hitherto unappreciated challenges faced by the strategy of confirmational holism. A properly informed analysis of scientific explanation is again required in order to see whether scientific practice can underwrite a naturalistic argument to a metaphysical conclusion (viz., mathematical realism).

6. CODA

Swoyer, a keen advocate of explanationism in metaphysics, engages in a contemporary debate on abstract objects with Dorr (Sider et al. 2008). The two debaters largely agree on the methodology of their domain of enquiry: metaphysical theories are to be evaluated on their explanatory merits, where those merits are judged like the explanatory merits of scientific theories. In general, Dorr grants,

> There is something very attractive about the idea that we should try to make progress in philosophy by learning from the disciplines in which progress is most manifest, namely the sciences. More specifically, *the proposal is that we should take as our starting point the large and impressive body of case-by-case epistemological judgments shared by all scientific realists*. We then decide what we ought to believe about controversial philosophical questions in accordance with whichever epistemological theory does the best job of accounting for and systematizing these data. (2008, 43; my emphasis)[42]

I agree. But as far as there are any judgments shared by all scientific realists, these judgments raise considerable challenges to explanationism in metaphysics. Inference

[42] This is how Dorr understands naturalism.

to the best explanation in metaphysics is badly in need of justification that is properly grounded in our best understanding of the nature of scientific theorizing and explanation, and the role and limits of inference to the best explanation in science.

ACKNOWLEDGMENTS

I received very helpful feedback from Anjan Chakravartty, Chris Daly, Steven French, Markus Schrenk, Matthew Slater, Laurie Paul, and others.

REFERENCES

Armstrong, David. 1983. *What is a Law of Nature?* Cambridge: Cambridge University Press.

Armstrong, David. 1997. *A World of States of Affairs*. Cambridge: Cambridge University Press.

Baker, Alan. 2005. "Are There Genuine Mathematical Explanations of Physical Phenomena?" *Mind* 114: 223–38.

Baker, Alan. 2009. "Mathematical Explanation in Science." *British Journal for the Philosophy of Science* 60 (3): 611–33.

Barnes, Eric. 1992. "Explanatory Unification and Scientific Understanding." In *PSA: Proceedings of the Biennial Meeting of the Philosophy of Science Association*, vol. 1. 1992, 3–12.

Bechtel, William, and Adele Abrahamsen 2005. "Explanation: The Mechanist Alternative." *Studies in History and Philosophy of the Biological and Biomedical Sciences* 36: 421–41.

Bigelow, John. 2010. "Quine, Mereology, and Inference to the Best Explanation." *Logique et Analyse* 53 (212): 465–82.

Bigelow, John, and Robert Pargetter. 1990. *Science and Necessity*. Cambridge: Cambridge University Press.

Biggs, Stephen. 2011. "Abduction and Modality." *Philosophy and Phenomenological Research* 83 (2): 283–326.

Colyvan, Mark. 2006. "Scientific Realism and Mathematical Nominalism: A Marriage Made in Hell." In *Rationality and Reality: Conversations with Alan Musgrave*, edited by C. Cheyne and J. Worrall, 225–37. Dordrecht: Springer.

Colyvan, Mark. 2012. "Road Work Ahead: Heavy Machinery on the Easy Road." *Mind* 121 (484): 1031–46.

Cordero, Alberto. 2011. "Scientific Realism and the Divide et Impera Strategy: The Ether Saga Revisited." *Philosophy of Science* 78 (5): 1120–30.

De Regt, Henk W. 2001. "Spacetime Visualisation and the Intelligibility of Physical Theories." *Studies In History and Philosophy of Science Part B: Studies In History and Philosophy of Modern Physics* 32 (2): 243–65.

De Regt, Henk W., and D. Dieks. 2005. "A Contextual Approach to Scientific Understanding." *Synthese* 144: 137–70.

Doppelt, Gerald. 2005. "Empirical Success or Explanatory Success: What Does Current Scientific Realism Need to Explain?" *Philosophy of Science* 72 (5): 1076–87.

Doppelt, Gerald. 2007. "Reconstructing Scientific Realism to Rebut the Pessimistic Meta-Induction." *Philosophy of Science* 74 (1): 96–118.

Dorato, Mauro. 2012. "Mathematical Biology and the Existence of Biological Laws." In *Probabilities, Laws, and Structures*, 109–21. Dordrecht: Springer.

Dorr, Cian. 2008. "There Are No Abstract Objects." In *Contemporary Debates in Metaphysics*, edited by T. Sider, J. Hawthorne, and D. Zimmerman, 32–64. Oxford: Blackwell.

Douglas, Heather E. 2009. "Reintroducing Prediction to Explanation." *Philosophy of Science* 76 (4): 444–63.

Ellis, Brian. 2009. *The Metaphysics of Scientific Realism*. Chesham: Acumen.

Field, Hartry. 1989. *Realism, Mathematics, and Modality*. Oxford: Blackwell.

Friedman, Michael. 1974. "Explanation and Scientific Understanding." *Journal of Philosophy* 71: 5–19.

Frost-Arnold, Greg. 2010. "The No-Miracles Argument for Realism: Inference to an Unacceptable Explanation." *Philosophy of Science* 77 (1): 35–58.

Gijsbers, Victor. 2007. "Why Unification Is Neither Necessary Nor Sufficient for Explanation." *Philosophy of Science* 74 (4): 481–500.

Grimm, Stephen. 2009. "Reliability and the Sense of Understanding." In *Scientific Understanding: Philosophical Perspectives*, edited by K. E. Henk, W. de Regt, and Sabina Leonelli, 83–99. Pittsburgh: University of Pittsburgh Press.

Harker, David. 2013. "How to Split a Theory: Defending Selective Realism and Convergence without Proximity." *The British Journal for the Philosophy of Science* 64 (1): 79–106.

Hawley, Katherine. 2006. "Science as a Guide to Metaphysics." *Synthese* 149: 451–70.

Karaca, Koray. 2012. "Kitcher's Explanatory Unification, Kaluza–Klein Theories, and the Normative Aspect of Higher Dimensional Unification in Physics." *The British Journal for the Philosophy of Science* 63 (2): 287–312.

Keller, Evelyn F. 2002. *Making Sense of Life: Explaining Biological Development with Models, Metaphors, and Machines*. Cambridge, MA: Harvard University Press.

Kitcher, Philip. 1989. "Explanatory Unification and the Causal Structure of the World." In *Scientific Explanation*, vol. 13 of *Minnesota Studies in the Philosophy of Science*, edited by P. Kitcher and W. C. Salmon, 410–505. Minneapolis: University of Minnesota Press.

Kuhn, Thomas S. 1977. "Concepts of Cause in the Development of Physics." In *The Essential Tension: Selected Studies in Scientific Tradition and Change*, 21–30. Chicago: Chicago University Press.

Kuipers, Thomas A. 2002. "Beauty, A Road to the Truth." *Synthese* 131: 291–328.

Ladyman, James. 2007. "Does Physics Answer Metaphysical Questions?" *Royal Institute of Philosophy Supplement* 82 (61): 179–201.

Ladyman, James. 2012. "Science, Metaphysics and Method." *Philosophical Studies* 160: 31–51.

Laudan, Larry. 1981. "A Confutation of Convergent Realism." *Philosophy of Science* 48: 19–49.

Lewis, David. 1986. *On the Plurality of Worlds*. Oxford: Blackwell.

Lipton, Peter. 1999. *Inference to the Best Explanation*. London: Routledge.

Lombrozo, Tanya. 2011. "The Instrumental Value of Explanations." *Philosophy Compass* 6 (8): 539–51.

Lyons, Timothy D. 2006. "Scientific Realism and the Stratagema de Divide et Impera." *British Journal for the Philosophy of Science* 57 (3): 537–60.

Maudlin, Tim. 1996. "On the Unification of Physics." *Journal of Philosophy* 93 (3): 129–44.

McLaughlin, Brian P. 2010. "Consciousness, Type Physicalism, and Inference to the Best Explanation." *Philosophical Issues* 20 (1): 266–304.

Morrison, Margaret. 2000. *Unifying Scientific Theories*. Cambridge: Cambridge University Press.
Norton, John D. 2003. "A Material Theory of Induction." *Philosophy of Science* 70: 647–70.
Okasha, Samir. 2000. "Van Fraassen's Critique of Inference to the Best Explanations." *Studies In History and Philosophy of Science* 31 (4): 691–710.
Pargetter, Robert. 1984. "The Scientific Inference to Other Minds." *Australasian Journal of Philosophy* 62 (2): 158–63.
Paul, L. A. 2012. "Metaphysics as Modeling: The Handmaiden's Tale." *Philosophical Studies* 160: 1–29.
Psillos, Stathis. 1999. *Scientific Realism: How Science Tracks Truth*. London: Routledge.
Psillos, Stathis. 2005. "Scientific Realism and Metaphysics." *Ratio* 18: 385–404.
Psillos, Stathis. 2011. "The Scope and Limits of the No-Miracles Arguments." In *The Philosophy of Science in a European Perspective*, edited by F. Stadler, vol. 2. Dordrecht: Springer.
Psillos, Stathis. 2012. "Anti-Nominalistic Scientific Realism: A Defence." In *Properties, Powers and Structures: Issues in the Metaphysics of Realism*, edited by A. Bird, B. Ellis, and H. Sankey, 53–69. New York: Routledge.
Putnam, Hilary. 1975. *Mathematics, Matter, and Method*. Vol. 1. Cambridge: Cambridge University Press.
Quine, W. V. O. 1995. *From Stimulus to Science*. Cambridge, MA: Harvard University Press.
Rogeberg, Ole. 2004. "Taking Absurd Theories Seriously: Economics and the Case of Rational Addiction Theories." *Philosophy of Science* 71 (3): 263–85.
Saatsi, Juha. 2009. "Form vs. Content-Driven Arguments for Realism." In *New Waves in Philosophy of Science*, edited by P. Magnus and J. Busch, 8–28. Basingstoke: Palgrave.
Saatsi, Juha. 2012. "Scientific Realism and Historical Evidence: Shortcomings of the Current State of Debate." In *EPSA Philosophy of Science: Amsterdam 2009*, edited by S. O. S. de Regt and Henk W. Hartmann, 329–40. Dordrecht: Springer.
Saatsi, Juha. 2016a. "On the 'Indispensable Explanatory Role' of Mathematics." *Mind*. Advance access.
Saatsi, Juha. 2016b. Historical inductions, old and new', *Synthese*, First Online.
Sellars, Wilfrid. (1963). *Science, Perception and Reality*. Atascadero, CA: Ridgeview Publishing Company.
Sider, Theodore. 2007. Parthood. *The Philosophical Review* 116 (1): 51–91.
Sider, Theodore. 2011. *Writing the Book of the World*. Oxford: Oxford University Press.
Sider, T., J. Hawthorne, and D. Zimmerman. 2008. *Contemporary Debates in Metaphysics*. Oxford: Blackwell.
Stanford, Kyle P. 2003. "No Refuge for Realism: Selective Confirmation and the History of Science." *Philosophy of Science* 70 (5): 913–25.
Stanford, Kyle P. 2006. *Exceeding Our Grasp*. Oxford: Oxford University Press.
Strevens, Michael. 2008. *Depth: An Account of Scientific Explanation*. Cambridge, MA: Harvard University Press.
Sturgeon, Nicholas. 2006. "Moral Explanations Defended." In *Contemporary Debates in Moral Theory*, edited by J. Dreier, vol. 6, 241–26. Oxford: Wiley-Blackwell.
Swoyer, Chris. 1999. "Explanation and Inference in Metaphysics." *Midwest Studies in Philosophy* 23: 100–131.
Swoyer, Chris. 2008. "Abstract Entities." In *Contemporary Debates in Metaphysics*, edited by T. Sider, J. Hawthorne, and D. W. Zimmerman, 11–31. Oxford: Blackwell.

Thomson, William. 1889. *Popular Lectures and Addresses*. Vol. 1, 2nd ed. Nature Series. London: Macmillan.

Thompson, Silvanus P. 1910. *The Life of William Thomson, Baron Kelvin of Largs*. London: Macmillan.

Trout, J. D. 2007. "The Psychology of Scientific Explanation." *Philosophy Compass* 2/3: 564–91.

van Fraassen, Bas. 1989. *Laws and Symmetry*. Oxford: Oxford University Press.

Vickers, Peter. 2013. "A Confrontation of Convergent Realism." *Philosophy of Science* 80 (2): 189–211.

Woodward, James. 2003. *Making Things Happen: A Theory of Causal Explanation*. Oxford: Oxford University Press.

Woodward, James. 2010. "Causation in Biology: Stability, Specificity, and the Choice of Levels of Explanation." *Biology and Philosophy* 25 (3): 287–318.

Woodward, James, and Chris Hitchcock. 2003. "Explanatory Generalizations, Part 2: Plumbing Explanatory Depth." *Noûs* 37 (2): 181–99.

Ylikoski, Petri, and Jaakko Kuorikoski. 2010. "Dissecting Explanatory Power." *Philosophical Studies* 148: 201–19.

9

Interventionism and the Missing Metaphysics

A DIALOGUE

James Woodward

APOLOGETIC PREFACE

A number of philosophers with a metaphysical orientation have criticized *Making Thing Happen* for its failure to provide an account of the metaphysical foundations or grounds or truth-makers for causal and explanatory claims. I originally attempted to write an ordinary paper responding to these criticisms but found this to be a very difficult undertaking: I realized that I disagreed with my critics about so much that putting everything into an ordinary "linear" argument was impossible. I also realized that important elements in my disagreement had to do not just with the ideas of the critics but with the rhetorical devices and strategies for conversational control that they sometimes employ. This led me to the idea of writing a dialogue that attempts to highlight the latter as well as a former. What follows is, I readily admit, a caricature that makes no attempt to be fair or balanced. Many of the philosophers I know who are analytical metaphysicians do not share the affect and attitudes of my Professor Metafisico, and many are far more knowledgeable about science. I hope that readers will take the dialogue in the spirit in which it is intended—as an attempt to be provocative and to raise in a sharp, unnuanced way some questions that deserve more attention than they have hitherto received. These include issues about just what metaphysical grounding consists in or amounts to, why (or when or for what purposes) it is *required*, and how providing metaphysical foundations relates to providing scientific explanations of a more ordinary empirical sort and to methodological concerns that at least in the past were regarded as an important component of philosophy of science. In particular, I want to raise the question of whether it is somehow obligatory that all philosophers of science do the sort of metaphysics associated with providing grounds or truth-makers or (as I maintain and hope) there are kinds of inquiry in philosophy of science having to do with methodology and the interpretation of the content of the particular scientific theories that can be pursued independently of the kinds of concerns that animate analytical metaphysicians.

Other themes that I have tried to illustrate include the following:

1. Ambivalence about Science: Many contemporary metaphysicians express great respect for science and even claim that what they are doing is in various ways continuous with science (either in content or method or both). Yet at the same time they make substantive claims and employ methods that do not seem to be part of any scientific discipline, as in their frequent appeals to "intuition" or to notions like metaphysical explanation.[1] This willingness to work both sides of the fence is a great argumentative resource, but a bit frustrating to interlocutors.
2. Conversation Control. One of the most striking features of contemporary metaphysical discussion, at least to an outsider like me, is the use of rhetorical strategies or framing assumptions that both automatically render certain considerations irrelevant and automatically render other considerations central to the discussion, regardless of what you, the non-metaphysician, would like to be talking about. In effect, these devices are employed to *force* the conversation to be only about issues the metaphysician regards as important. One device for doing this is the adoption of highly expansive conceptions of "metaphysics" so that if, for example, you try to define anything or clarify a concept you are making a "metaphysical" claim. Similarly, ordinary empirical claims of some generality (e.g., about whether matter is continuous) are taken to be metaphysical claims.[2] Given that virtually everything is metaphysics (or at least that boundaries of "metaphysics" are very elastic and

[1] In addition, I think that a great deal of analytical metaphysics is animated by mistaken ideas about the methods actually employed in successful science. In particular I think that analytic metaphysicians tend to exaggerate the role of transempirical virtues and methods in science, including appeals to "simplicity" and "inference to the best explanation." However, this is a topic for another paper.

[2] To clarify: I am not opposed to all expansive conceptions of metaphysics. If one engages in, say, some wide-ranging, empirically informed exploration of what nature is like, given our present scientific understanding of it, and calls that "metaphysics," far be it from me to object. What I object to is the use of such an expansive conception as a device to justify the characteristic preoccupations of analytic metaphysicians in the manner illustrated below.

ill-defined), it becomes difficult to argue that one is not doing metaphysics, and the non-metaphysician becomes vulnerable to Metafisico's rhetorical maneuvers (since you are already doing metaphysics, you might as well do it right, which requires a discussion of grounding or something similar). Another device is the invocation of various distinctions such as a contrast between those features of science that have "merely" epistemic or methodological or "pragmatic" significance, and those that have true ontological/metaphysical significance, with only the latter being treated as a fit subject for philosophical discussion. The language metaphysicians use to characterize their concerns (frequent use of words like "deep," "fundamental," "ontologically serious," and so on, with the apparent implication that non-metaphysical projects are superficial, shallow, and non-serious) has a similar rhetorical role in directing philosophical discussion, as does the use of phrases like "merely pragmatic" as terms of denigration. I will add that I am convinced that many metaphysicians are unaware that the rhetorical devices they employ look to those who are not metaphysicians as though they are tailor-made to achieve such exclusionary effects—instead they think of the devices as both natural and completely innocent. But words like "deep" and "fundamental," when used in contexts of this sort, are not innocent; they reflect the interests and tastes of metaphysicians disguised as something more "objective."

3. Preferences and Arguments. Related to (2), expressions of subjective preferences for (or value judgments regarding) certain kinds of philosophical work or topics presented as though they are "objective" arguments that require that one work only on these topics. From my point of view, it would be very refreshing if more metaphysicians would just say that they find doing the metaphysics of grounding more interesting or personally engaging than, say, trying to understand the role of experimentation in causal inference and that for this reason they prefer to work on the former. Instead, what we too often get are claims that work on the former is logically required if one works on the latter, that the former is "deeper," and so on.
4. Details: If you think that you already know that some philosophical position of a certain general type *must* (for metaphysical reasons) be correct, you may not be too concerned about working out the details of that position. Apparent problems can be dismissed because (since your position is correct) you know it must be possible in principle to meet those objections. Many defenses of one of the main topics of this dialogue—the Best Systems Analysis—seem to me to have this character.
5. Methodology. There are many issues and problems, arising in the various sciences, that are, broadly speaking, epistemological or methodological in

nature. These have long been regarded as legitimate and important topics for philosophical discussion and, as I see it, are central to philosophy of science. They include, for example, investigation (both descriptive and normative) of different strategies for learning about nature (including investigations of the limits of what can be learned from various sorts of evidence), and the characterization and evaluation of various forms of reasoning that figure in different areas of scientific and non-scientific investigation. Work of this sort is pursued not only in philosophy of science, but also in statistics, machine learning, discussions of experimental design, artificial intelligence, and portions of cognitive psychology, not to mention the various sciences themselves. I would also include under this general heading of methodology efforts at clarification of concepts that are central to the various particular sciences, as in Kendler's (2005) discussion of what it is for a gene to "cause" a phenotypic trait. Analytical metaphysics and allied work in the metaphysics of science seems to me to take little notice of methodological issues and concerns of the sort just described. This is partly a consequence of the tendency of contemporary metaphysicians to downplay or attempt to abstract away from epistemic and "pragmatic" considerations, which makes it difficult to discuss anything relevant to methodology. What follows is in part an attempt to push back against this tendency and to argue for the independent value of methodological concerns.

THE DIALOGUE

JW: Hi. I'm Jim Woodward. Pleased to meet you. And your name?

M: I am Professor Metafisico, PhD.

JW: What do you work on?

M: I'm an Analytical Metaphysician.

JW: I'm a philosopher of science who works on causation, among other things.

M: That sounds interesting. I'd like to hear the details.

JW: I try to develop the idea that causal claims describe relationships that are potentially exploitable for purposes of manipulation and control. Put slightly differently, causal claims describe the outcome of hypothetical experiments: very roughly, *X* causes *Y* if and only if there is some intervention *I* that changes the value of *X* such that if *I* were to occur, the value of *Y* would change. An intervention on *X* with respect to *Y* can be thought of as an idealized, unconfounded experimental manipulation of *X*. When the relationship between *X* and *Y* is a causal relationship, it will be stable or continue to hold—*invariant*—under at least some interventions on *X*. Causal relationships differ in their degree of invariance, with some causal relationships being invariant under a wider range of interventions and other sorts of changes than others. There is more detail in my book, *Making Things Happen* (*MTH*).

M: Hold it right there. I hate to break the news to you, but your idea is a non-starter—it sounds like you're saying that whether *X* causes *Y metaphysically depends* on what would happen to *Y* under interventions on *X*. But that can't be right. The claim on the right hand side of your definition is a counterfactual. Counterfactuals can't be "barely true"—they require "truth-makers" or "grounding"[3] in terms of what is non-counterfactual (i.e., categorical) and more fundamental. "Truth supervenes on Being," as we metaphysicians like to say. Interventions and even more so claims about what would happen under interventions are not plausible candidates for what is Ontologically Fundamental.

JW: Actually, the above "definition" or bi-conditional was not intended as a claim about metaphysical dependence or anything like that. I put it forward because I thought it captured the way that causal claims are used or understood in a number of areas of science—especially but not exclusively the social behavioral and biological sciences—and in common sense reasoning about causation and because I thought this in turn had interesting methodological implications. I claimed in *MTH* that this way of thinking about causal claims was fruitful because it helped to clarify what researchers in the above disciplines mean or are committed to (or are trying to establish) when they make causal claims and also how such claims relate to ideas about causation advanced by philosophers.

I also suggested that interpreting causal claims in this way had a normative or regulative significance: vague or unclear causal claims may be disambiguated by spelling out the hypothetical experiments with which they are associated. A related point is that we can often clarify the sort of evidence and

[3] There are a number of different positions and accompanying terminology that are in the general neighborhood of Metafisico's views. Some writers talk about "grounding" causal and other sorts of claims, others about identifying the "truth-makers" for such claims or about specifying what such claims "metaphysically (or ontologically) depend on." While acknowledging these differences, in an effort to make this dialogue readable, I have deliberately lumped these different positions together. So for "grounds" the reader should read: "grounds or truth conditions or . . ."

other assumptions needed to establish a causal claim on the basis of non-experimental data by reflecting that this must be evidence that is sufficient to establish what the outcome of the associated experiment would be, were we able to perform it. This helps us to see why and when certain techniques such as the use of instrumental variables in econometrics can be used to reliably establish causal conclusions. I have always thought of these goals as interpretive/descriptive/semantic/methodological (methodological for short) rather than metaphysical—somewhat similar in aspiration to the treatment of causation in Holland (1986) or in King, Keohane, and Verba (1994) or Morgan and Winship (2007) and having some affinities with and drawing on ideas from Spirties, Glymour, and Scheines, (2009), and Pearl (2009). Many of these writers also offer what they call "definitions" of various causal notions, but this is more in the service of the methodological goals I mentioned and not because they are interested in making claims about "metaphysical dependence." In fact, definitions abound in science and mathematics, but they are rarely if ever intended as claims about metaphysical dependence—books on real analysis typically define the continuity of a function in terms of the notion of an open neighborhood, but I do not think they mean to claim that continuity metaphysically depends on this notion. I will add, though, that from my point of view nothing really turns on my use of the word "definition"—I am happy to drop it in favor of talk of relationships of mutual constraint between causal claims and claims about the outcomes of hypothetical experiments or in favor of talk about using the latter to elucidate or clarify the former with an eye to the goals described above[4].

M: These people you describe above—I haven't heard of any of them. Where did they go to philosophy graduate school? Who did they study Metaphysics with?

JW: They don't have PhDs in philosophy and they aren't aiming to do metaphysics. Holland is a statistician, King a political scientist, and Pearl a computer scientist.

X: Well, I don't mean to be dismissive but it seems unlikely in that case that they will be able to make any contributions to the Fundamental Ontology of causation. It sounds like what they are doing is at best of merely practical or pragmatic interest.

JW: I'm OK with that. More than OK in fact. I think it is worthwhile to do things that are of merely pragmatic interest. That is precisely my point. They

[4] Some of the ideas described in this paragraph are defended in more detail in Woodward 2015.

aren't trying to make contributions to Metaphysics. They are trying to do something else, and similarly for me.

M: Well, the work you describe may be of interest to some people outside of philosophy in its unambitious way, but it doesn't sound as though what either you or they are proposing has any "metaphysical depth" to it.[5] I fear that it lacks what John Heil (2003, 2), echoing various Australian philosophers, calls "ontological seriousness."

JW: Once again, I'm happy to agree. It seems to me that the people I mention just have a different set of interests from you. And the same is true for me. Why can't we just agree on a division of labor? You can do metaphysics and I'll do what interests me.

M: What you call a division of labor is not possible. By *definition* metaphysics is concerned with "the foundations of reality" (Chalmers, Manley, and Wassserman 2009, 1), with "Being as such" (van Inwagen 2007), and with "the fundamental structure of the world" (Sider 2009, 420). Everything is thus suffused with or at least grounded in metaphysics and, as a result, you can't *not* do metaphysics (Sider 2009, 420). Indeed, if you deny you are doing metaphysics, you are by that very denial doing metaphysics since you must have in mind some conception of what it is that you are not doing and you must think that it is possible not to do that. This itself amounts to a metaphysical position that needs to be defended—by doing metaphysics of course. Oh, and I almost forgot, if you discuss the ideas of anyone who is a metaphysician you also must be doing metaphysics (Strevens 2007).

JW: It sounds as you are saying that I have to do what you are doing, according to your rules. And you don't have to provide any motivation or justification for what you are doing, since one can't not do it.

M: Your description of the dialectical situation is essentially correct, although rather unsympathetically expressed. Let me emphasize, though, that these requirements concerning the centrality and inescapability of metaphysics do not in any way reflect the personal ambitions of us metaphysicians. It is not as though we are trying to enhance our own careers when we insist on these requirements. Instead, the requirements flow ineluctably from the foundational role of metaphysics in all inquiry. And since, as we have established, you can't avoid doing metaphysics, you might as well buckle down and do it thoroughly and correctly, even if that means that you never get around to doing any

[5] Approximately quoting a prominent philosopher of mind commenting on my views at a recent conference. He didn't intend this as a compliment, but I was happy to agree.

of this stuff you call methodology (aside: which wouldn't be a bad thing at all, in my opinion).

JW: I would find it easier to understand what you are getting at if you would stop using words like "depth" and "serious" and explain to me exactly what is wrong with being interested in the merely methodological. For example, will I make mistakes in the methodological projects that interest me if I fail to provide grounds/truth-makers in your sense for causal and counterfactual claims? Will this lead to empirical mistakes of some kind?

M: How can any area of inquiry be satisfactory if it is left ungrounded and without foundations? As for specific problems that arise for your project if you fail to provide metaphysical foundations, I haven't worked out the details since I don't think this is necessary, but I think it is obvious that you'll end up with an account that is unclear, incomplete, and subject to other problems as well. And of course it won't be Ontologically Serious and won't be a contribution to anything that is Fundamental.

JW: But the people mentioned above don't provide metaphysical foundations and yet what they say doesn't *seem* to be confused. Or if what they say is defective in some way, this doesn't seem to be because they have failed to provide metaphysical foundations. Can you point to specific problems that arise with their accounts because they fail to provide metaphysical foundations? For example, Rubin (1974) and Holland (1986) employ what is in effect a counterfactual treatment of causation, defining the causal effect of a treatment t with respect to an alternative treatment t' for a unit u as $Y_t(u) - Y_{t'}(u)$ where $Y_t(u)$ is the value Y would have assumed for unit u if it had been assigned treatment t and $Y_{t'}(u)$ is the value Y would have assumed had u instead been assigned treatment t'. This is then used to characterize various sorts of assumptions that may be used for reliable inference to causal conclusions from statistical data—the idea is that the assumptions must be such that in conjunction with the statistical data, they allow you to estimate $Y_t(u) - Y_{t'}(u)$ or perhaps the expectation of this quantity $E[(Y_t(u)) - (Y_{t'}(u)])$. These ideas are criticized by some, but these criticisms don't seem to have much to do with any failure on Holland's part to provide a proper *metaphysics* of causation. Pearl, for example, argues that it is often easier and more straightforward (and we are less likely to make mistakes) if we employ graphical methods rather than variables describing counterfactual outcomes like $Y_t(u)$, but Pearl's graphical representations don't supply truth-makers in the sense that you seem to have in mind either, and in any case Pearl doesn't seem to be claiming that they are metaphysically superior to treatments like Holland's. My own view is that the notion of an intervention provides a more precise clarification of Rubin and Holland's notion of the "assignment" of a treatment to a unit—a notion that is left underspecified in their accounts—but again this doesn't seem to turn on issues having to with metaphysical foundations.

M: It does sound from your description as though the interests of these people are, sadly, merely pragmatic. But exactly because of this, I don't see that what they are doing is of any Metaphysical Interest. You still haven't told me anything about what the truth-makers or grounds are for causal claims.

JW: I thought that we were talking about your claim that various problems will arise for what they (and I) are doing if we fail to provide metaphysical foundations. You seem to be confusing this claim with the assumption that they ought to be doing metaphysics and/or that only metaphysics is worth doing. This assumption amounts to a *preference* on your part for certain kinds of inquiry. You seem to be confusing this preference with an *argument* that what you advocate doing is required or unavoidable.

M: Let me give it another try.[6] A while back you asserted a bi-conditional linking causal claims to the outcomes of hypothetical experiments. And you also described what you were doing as an attempt to capture what those who use causal claims in some areas of science "mean" or are "committed to" in connection with those claims and that causal claims can be clarified by associating them with hypothetical experiments. But everyone agrees that when you assert such a bi-conditional, and particularly when you connect it to notions like "meaning," you are making a metaphysical assertion—you are making a claim about the grounds or truth-makers (or something similar) for causal claims. So I don't understand how you can claim that you are not trying to do metaphysics.

JW: I don't understand why the claims that I make about causation *must* be interpreted in the way you describe. You seem to be treating this as a forced move of some kind that I am required to make. As I said above, I'm not particularly wedded to describing what I'm doing in terms of providing definitions or meanings, especially if you are going to then go on to interpret what I'm saying in the metaphysical way you want to foist on me. Perhaps it would be better (at least for the purposes of this conversation) if I put what I am saying in terms of characterizing a certain set of practices for reasoning about causal claims, connecting them up with other sorts of claims, making distinctions among such claims, and so on. Describing and elucidating aspects of the

[6] Thanks to Martin Thomson-Jones for raising some issues connected to what follows.

behavior of certain concepts (like "cause" and "intervention") and connecting them to methodological goals, or something along those lines. Surely there are activities of this sort that go on all the time in science and among many who comment on science (both philosophers of science and others), but which we don't have to view as bound up with metaphysics in the way you have in mind. You seem to be simply stipulating a notion for "meaning" and, for that matter, "if and only if" that builds in the idea that one can't do what I am trying to do without making heavy-duty metaphysical claims about grounding. Why can't I just decline to accept your stipulation?

That said, I wonder if it wouldn't be more fruitful to try another tack. Perhaps you might tell me more about this grounding or truth-making relationship that I'm not adequately dealing with. That way I would have a better sense of what I am leaving out.

M: I'd be delighted to—this is the subject of a rich and flourishing literature (see, e.g., Chalmers, Manley, and Wasserman 2009; and Correia and Schnieder 2012).

Let me begin with grounding, although I should acknowledge at the outset that there is a fair amount of disagreement about just what this relationship involves or even whether it is intelligible and about whether it is or is not the same thing as truth-making. For starters, many think that the grounding relation is primitive, and thus it can't really be explained in terms of anything else (Schaffer 2009). But I can tell you about some of its properties. For example, many metaphysicians hold that it is irreflexive, asymmetric, and transitive. Grounding and the metaphysical definitions in which it figures are distinguished from more ordinary sorts of definitional relationships, which may permit you some choice about which terms one starts off with and which terms are defined in terms of them, as is often the case with various alternative sets of definitions and axiomatizations in logic and mathematics. (Think of various choices of primitives in propositional logic or in Euclidean geometry.) By contrast, in the case of proper metaphysical definitions that express relations of metaphysical dependence, there is a uniquely correct order of definition : p is defined in terms of its grounds or truth conditions q, q is defined in terms of its grounding conditions, and so on until we reach those entities and relations that are fundamental. Grounding relations thus give us insight into the hierarchical structure of reality, according to which some entities or facts are dependent for their existence on others, while other entities or facts are fundamental, requiring no further grounding. Interestingly, there is a striking parallel here with the social organization of philosophy: just as Philosophy programs (and individual philosophers) should be ranked in a hierarchy, according to their Excellence in Metaphysics, so also Reality itself can be ranked, in terms of its relation to what is Fundamental. To be is to be Ranked.[7]

JW: Fascinating. Tell me more.

M: The grounding relation has other properties as well. For example, it is hyper-intensional, so that logically equivalent claims can have different grounds. The grounding relationship is also related to the best theory we have of everything—the best theory should be thought of as a theory of what is fundamental that correctly tracks grounding relations (Schaffer 2009). Among scientifically minded philosophers, it is assumed that the best candidate for such a theory is to be found in Fundamental Physics—a view to which I myself subscribe. So the grounding entities (or whatevers) will presumably include, for example, elementary particles and fields of the sort discussed in fundamental physics. Fortunately, though, we needn't inquire too closely into the details of this—we can take this aspect of what is fundamental just to have to do with "point particles and fundamental physical magnitudes" (Schaffer 2009, 370) and leave the rest to the physicists. So to sum up, when we metaphysicians ask for the grounds or truth makers for causal claims or claims about laws, what we are asking for are facts or entities that stand in the kind of relationship to causal claims and laws that is common to all of the examples described above.

JW: I'm still a little confused. On the one hand, it sounds as though metaphysics and grounding, as you describe them, are (at least among scientifically minded metaphysicians) closely connected to physics—apparently to what you think of as fundamental physics. On the other hand, I take it that metaphysics is not exactly the same thing as physics? So what exactly is their relation?

M: Obviously metaphysics is not the same thing as physics. If it were, what would be left for metaphysicians to do? The very fact that metaphysicians keep doing metaphysics, as they have done for thousands of years, and do it, at present, in a way that has no influence whatsoever on the practice of physics, shows that what we are doing must be importantly distinct from physics and the other sciences and valuable in its own right. As for the relation between these disciplines, this is a matter of on-going discussion. One possible view is that metaphysics is inspired in part by physics, but must satisfy other constraints too—for example, consistency with the intuitions of metaphysicians about grounding. Physicists, although admirable in many ways, tend to be inattentive to the need for grounds. For example, they may introduce claims about

---

[7] Or, in a more "continental" mode, Being Ranks Itself.

laws of nature, and support these only with empirical evidence, and physical arguments about symmetries and so on, while failing to specify what the ultimate grounds for these law-claims are. One useful thing that metaphysicians can do is to step in and provide more adequate grounds for these claims. It is your failure to do this that makes your views about causation so defective and lamentable.

Another possible view of the relation between science and metaphysics (perhaps not sharply distinct from the view just described) is that metaphysicians provide accounts of the basic categories in terms of which all scientific theories must be framed—categories like "cause," "law," "substance," "part" and "whole," and so on. Scientists then fill in the details, providing particular examples of causal relationships, laws, and so on. So in this picture as well, metaphysics plays a foundational role in science.

Yet another other thing that distinguishes us metaphysicians from what physicists do is that we are not so interested in, so to speak, actually existing physics, but rather in the wider space of metaphysically possible worlds. Our aim is to discover what the grounds would have to be for any metaphysically possible physics.[8] This is why we are concerned to make our theories physics-like, or physics-inspired, rather than being too closely tied to the physical details of our world, since the latter course would lead to a regrettable loss of generality. For example, in discussing laws of nature, we find it most useful to think of these just as generalizations of the form "All *A*s are *B*s" and in discussing scientific theories we represent them as theory *T*. Any further detail would just be limiting.

Perhaps I should add that as a scientifically oriented metaphysician, I fully accept Quine's dictum that metaphysics is continuous with and "on a par" with science and that metaphysicians are just doing at a somewhat more abstract and general level what those physicists at CERN who were looking for the Higgs boson were doing. The criteria for a good metaphysical theory are at a general level the same as those for a good scientific theory, simplicity, great scope, theoretical fertility, and so on, as many metaphysicians have said (e.g., Sider 2009). On the other hand, it also needs to be borne in mind that we are doing something that is more fundamental than any particular science and that shouldn't be judged just by the sorts of pragmatic criteria that are unfortunately sometimes employed in the assessment of scientific theories, like usefulness in prediction and control, computational tractability, and so on.

---

[8] Cf. Lowe (2011, 100), who characterizes metaphysics "as *the science of the possible*, charged with charting the domain of objective or real possibility."

A good way of summing up is that what we are doing is both science-like and not science-like. I hope you find this clarifying.

JW: It would help me if you could provide some more specific examples of grounding.

M: I'm happy to oblige. The literature is full of illustrations: the disjuncts of a disjunction ground the disjunction; geology is grounded in physics; according to some philosophers, causal claims are grounded in facts about dispositions; according to other philosophers, causal claims are grounded in facts about laws of nature; and laws of nature are grounded in relations of necessitation between universals. According to still other philosophers, laws of nature are instead grounded in facts about the deductive systemization of the entire Humean Basis that achieves a best balance of simplicity and strength. Other illustrations: fists (if they exist—there is dispute among ontologists about this) are grounded in hands or shapes assumed by hands, and the set whose sole member is the empty set is grounded by the empty set. However, I caution that these are just examples. The grounding relation is the single unique relation that is common to all of these particular cases. Thus when we metaphysicians ask for the grounds or truth makers for causal claims or claims about laws, what we are asking for are facts or entities that stand in the relationship to causal claims and laws that is exemplified in all of the above examples.

JW: I'm afraid I'm still not getting it. Why do you think that these are all common exemplifications of a single grounding relationship? It looks to me that in the first example, what you have is an asymmetry of logical entailment between a disjunction and its disjuncts. In the second case, you are appealing to some idea according to which the entities that are the subject matter of geology are made up of entities like atoms that are studied in physics. So in this case you are talking about some sort of compositional relation. Or perhaps you are claiming the geological phenomena can be explained in terms of (or "reduced to") relationships that figure in physics, where "explanation" here means scientific or physical explanation. In still other cases, your grounding relations seem to be conceptual or definitional in character. At any event, when you describe all of these cases in terms of grounding or truth-making, don't you risk conflating causal or explanatory dependence (of the sort discovered in empirical science), with logical entailment, with part/whole relations, with various sorts of semantic or definitional relations, with . . . ? I've always thought that one of the great achievements of early modern era in philosophy was the clear separation of these notions, which tended to be conflated in earlier scholastic thought. Indeed, don't locutions like "truth-maker" encourage such conflation, since "making" sounds like a causal notion, but at the same time truth-making relations seem to be conceptualized as something very different (perhaps as involving logical or conceptual relations of some kind) from ordinary causal relationships? Perhaps you should consider the possibility that there is no single relationship that has all of the features of grounding or truth-making, at least as you are conceiving of it.

M: I don't see how that is possible. Grounding and truth-making have been central to metaphysics since Aristotle, if not Thales, and discussion of these continues in the Best Departments today. How could all of these people be wrong about whether there is such a relationship? And if there is a grounding relationship, there could hardly be more than one, could there? I mean, what would happen to Fundamental Ontology (not to speak of our Departmental Ranking) then? I would say that the very fact that metaphysicians talk (and disagree) about grounding (and/or truth-making) shows that these notions (or at least one of them) must be intelligible and that is our job to discover its properties.

JW: I see. Let me then try to express a somewhat different worry. You agree that there is considerable disagreement among metaphysicians about what grounding (or for that matter related notions) involves and how to understand this notion.

M: Oh, yes. This is a very lively area of debate, with a wide spectrum of different views.

JW: But then perhaps you can also see that this puts me at a considerable rhetorical disadvantage when metaphysicians like you demand grounds for causal claims and counterfactuals. I mean, if you cannot agree among yourselves about what you are talking about when you ask about grounds, it is going to be difficult for someone like me to respond to your demands.

M: That is your problem, not ours. I can tell you that it probably will take some time for us metaphysicians to work out the full theory of grounding. And perhaps, alternatively, we will decide that talk of grounding should be replaced by something else (Truth-making, Structure, or what have you). But we are all agreed that grounding (or something like it) is crucially important, and it is a great failure on your part that you have not provided a detailed story about how this works in connection with causal claims and counterfactuals, even if we have not yet fully decided what grounding is.

JW: Perhaps we are more likely to get somewhere if we talk more about specific examples. Consider the behavior of a macroscopic sample of dilute gas conforming to the ideal gas law and other familiar generalizations of thermodynamics.

In this case, would the underlying "grounds" for this behavior consist in, say, facts about its constituent molecules and the laws governing their interactions of the sort that would be described in statistical mechanics?

M: Something like that. As I've already suggested, we metaphysicians largely leave the details of which grounding relationships are actually exemplified in our world to physicists, except when they make obvious metaphysical mistakes. Presumably the true grounds in the example you describe involve far more fundamental physics—truths of string theory or something similar. In any case, the details don't matter—it is the general idea of grounding that is philosophically important. This is why merely providing this underlying physics does not by itself answer the metaphysician's concerns about the proper elucidation of the grounding relationship.

JW: Well, here is something that puzzles me. Many, many different assumptions about the micro-constituents of matter and their interactions are consistent with the holding of the ideal gas law—even the assumption that the gas is a continuum is consistent with this law, given the right general assumptions about the behavior of this continuum. The same is true of many other aspects of the behavior of macroscopic objects—their behavior is surprisingly independent of the details of the underlying physics, often depending only on very generic features of that physics. For example, aspects of the behavior of many substances of quite different material composition undergoing phase transitions depend only on very general features of those systems—their dimensionality, the symmetry properties of the Hamiltonians governing these systems, and so on. This is why, to paraphrase Nigel Goldenfeld and Leo Kadanoff (1999), to model a bulldozer you don't have to model its constituent quarks. So my question is why aren't the grounds or truth conditions (or whatever) for this behavior something like the following: either the sorts of generic features just described or perhaps a disjunction or equivalence class of all theories consistent with the macroscopic behavior, rather than whatever highly specific "low-level" physical truths about strings or whatever hold in the actual world? After all, in one obvious sense the macroscopic behavior does not depend on these very specific truths at the fundamental level. So if the grounds for the behavior of macroscopic entities have to do, among other things, with those features to which we appeal to provide ordinary scientific explanations of that behavior, it seems to me that there is a good case to be made that these features in many cases are not going to involve highly specific truths of some fundamental physical theory. Again, perhaps this is a reflection of the fact that you are trying to do too many things at once with your grounding/truth-making idea—these are supposed to reflect underlying physics, provide explanations, provide reductions to supposedly conceptually more primitive facts and much more, but maybe a single relationship can't embody all of these features.

M: I haven't really thought about the issues you describe about the independence of the behavior of macroscopic objects from the physics that underlies them, and, frankly, they involve details of a sort that bore me. But to the extent you are suggesting that the grounds in such cases are generic or disjunctive facts, your proposal is a non-starter. Really, you seem to have no ear for metaphysics at all. I've already emphasized that disjunctions cannot be metaphysically fundamental but must rather be grounded in their disjuncts. Moreover, it would be contrary to all Metaphysical Intuition if something as abstract and generic as the symmetry properties of a Hamiltonian could serve as a ground. In general, the difficulty of modeling a bulldozer by means of quantum chromodynamics is merely pragmatic and of no philosophical significance. When we put modeling considerations aside, we see that the truth-makers for the behavior of bulldozers indeed are to be found in the behavior of its constituent quarks (or something like that).

JW: Part of the reason why these issues interest me is that it seems to me that one of the great advantages of experimentation, epistemically and methodologically speaking, is that one can often use it to reliably establish conclusions about causal relationships independently of underlying details that would be provided by some lower-level theory. The interventionist account attempts to capture this observation. For example, a researcher may be able to establish that some drug provides a cure for an illness without knowing the underlying chemistry of the drug, much less how its action would be modeled in quantum field theory. This is another case of what I was describing above: a kind of independence or decoupling (in this case epistemic independence) of more macroscopic behavior from underlying physical details. From the point of view of methodology and discovery, the most reliable sources of knowledge about the behavior of more macroscopic systems often is *not* found in information about what you are calling grounds or truth makers, insofar as these have to do with fundamental physics. Perhaps that is part of the reason I don't find a concern with grounding, even if intelligible, particularly relevant to my interests.

M: I regard these observations as completely irrelevant. Indeed, I'm shocked that you would allow your philosophical views about causation or anything else to be influenced in this way by merely epistemic considerations. This is just vulgar. The whole point of our discipline is to get behind the merely practical and epistemic and to limn the True Nature of Reality.

JW: OK. Let me try something different. Consider the counterfactual claim that if right now I were to release this wine glass I am now holding in my hand, it would fall to the floor. If you want grounds or truth-makers for this claim, why can't I just say that these have to do with whatever would figure in a deeper physical explanation of why the counterfactual claim is true—for example, there is the gravitational force exerted on the glass by the earth in accordance with the inverse square law, the fact that, as it falls, the resisting force of the surrounding air, modeled by Stokes law, will be negligible, for an object of the dimensions of the glass, the initial conditions of the system once the glass is released and . . .

M: I'm afraid that once again you're completely missing the point. What you have just described is a mere physical explanation[9] and just a particular example of one at that. To repeat: Metaphysicians are after something different and deeper—the metaphysics in virtue of which your counterfactual is true. You can supply all the physics you want and you still won't have given me the metaphysics that is so lamentably missing from your account. What we are looking for are the metaphysical facts and relationships that underlie and make true all lawful or causal relationships rather than specific examples of such relationships. (At the same time, though, let me remind you, as emphasized above, that what we are doing is continuous with and "on a par with" science.) In the case you mention, for example, a metaphysician will want to know what grounds or serves as a truth condition for the inverse square law to which you appealed in your physical explanation. Surely you can see that it is not satisfactory to just take the inverse square law as metaphysically primitive.

JW: I wasn't suggesting that we take the inverse square law as metaphysically primitive, if only because I don't know what "metaphysically primitive" means. I was instead hoping we could find a way of talking about laws that did not get us embroiled in a discussion of what is metaphysically primitive or derivative. But perhaps we could make better progress if you would give me a specific example of what counts as a ground or metaphysical truth-maker for a causal or counterfactual or nomological claim. What have metaphysicians proposed along these lines?

M: I'm happy to oblige, although there are lots of candidates, since again this is an area in which there is a lot of disagreement. (A reflection, once more, of

[9] JW is compelled to add that in many cases, no one has any idea how to connect causal and counterfactual claims in the special sciences to the fundamental laws that "underlie" them. Because of this, it is opaque how to connect these laws to issues about the testing of such claims, or how to reason with them. Observations about the grounding of such claims in fundamental physical law, even if correct, do not seem to help with such issues.

the great intellectual richness of our subject.) Some philosophers hold that the truth-makers for laws (and presumably also for causal claims and counterfactuals) are to be found in special entities (or properties, or relationships). For example, according to Dretske (1977), Armstrong (1983), and Tooley (1987), the truth-makers or grounds for laws of nature are relations of necessitation between universals, and according to other philosophers (e.g., Bird 2007), laws are made true by facts about dispositions or "active powers" possessed by particular objects. I myself prefer a second kind of approach, the best systems analysis of laws (hereafter the BSA), which strikes me as more scientific in spirit and also promises a satisfactorily Humean reduction of the notion of law.

JW: I'd like to hear more about the BSA. I'm relieved to hear that you prefer it to the other alternatives you describe because frankly, I'm a more than a little unclear about exactly what they accomplish. Suppose I claim that the gravitational inverse square law or the Lorentz force law holds for some domain of investigation. What exactly have I added if I then supplement this with the claim that both are made true by a relation of necessitation between universals? Presumably this necessitation relation must possess features that exactly mimic whatever other features we think are possessed by the Lorentz force law. Moreover, we don't seem to have any evidence that the necessitation relation possesses those features that is distinct from whatever evidence we have that the Lorentz force law possesses those same features. For example, if we think that the Lorentz force law licenses certain counterfactuals, then we will also postulate that the associated relation of necessitation among universals also operates in such a way that it licenses the same counterfactuals. So it looks as though all I have done is to translate the law and whatever is implied by it into necessitation talk. Or suppose I say that every individual charged object has a disposition to conform to the Lorentz force law and this is what "grounds" the law. Doesn't this (at best) just re-describe the law in some new terminology? Worse, don't we have a sort of illusion of explanation: We claim that the inverse square law or at least the regularity it describes, holds "because" of a relation of necessitation between universals and so on. But I don't see in what sense we have an "explanation" here, rather than (at best) just a redescription.

M: Well, although I prefer the BSA, I think that, speaking as a metaphysician, you are being unfair to the views just described—they needn't be trivial in the way you suggest. For example, some versions of dispositional accounts make the important and distinctive claim that the truth-makers are located "in" individual objects in the form of dispositions they possess, as opposed to being located somewhere else that is external to those objects. Stephen Mumford (2004) uses this idea to argue that there are no such things as laws of nature, at least as ordinarily conceived. This is certainly a non-trivial claim that has generated a great deal of discussion. If correct, it shows that most physicists misunderstand their own discipline—a result that we metaphysicians would welcome, since it would demonstrate the great importance of our investigations.

As for your claim that the accounts are just re-description, I agree that they do not provide ordinary scientific or causal explanations. Instead they aim to provide a special kind of explanation—a "metaphysical explanation."

JW: I'm guessing that we wouldn't find it very profitable to discuss why what you call "metaphysical explanations" should be regarded as explanatory. I'll add that I would be inclined to regard it as a liability rather than a virtue of the search for truth-makers that it apparently can be used to generate questions about where the truth-makers for laws of nature are "located." But since we both agree that the special entities accounts are unpromising, why don't we move on to the BS approach?

M: Excellent. Here is the basic idea: Start with the full Humean Basis (HB)—a specification everything that actually happens throughout the entire lifetime of the universe, but purged of any references to modality, possibility, causal or lawful relations, or anything like that. Then consider alternative systemizations of this HB, looking for those that achieve a best balance of simplicity and strength. The laws will be those regularities that are described by the axioms and theorems that are common to all such best systemizations. Assessments of simplicity are to be made with reference to a canonical or privileged language that consists of predicates corresponding to the "perfectly natural properties."

JW: And why exactly should we believe that laws are connected in this way to finding a best balance of simplicity and strength?

M: The BSA has a clear answer to this question: these standards—simplicity, strength and achieving a best balance between them—are the standards (in fact, the only standards) that scientists actually use in choosing among competing theories. And when they are applied to the problem of laws, they yield an account with all sorts of attractive features. For example, the resulting account satisfies our aspiration for a Humean reduction of claims about laws to claims about regularities (or at least to claims about the way that regularities fit together in a total deductive system).

JW: So you are saying that these are the standards that are actually employed in scientific practice in choosing among alternative theories and that when one chooses in accord with them, one picks out exactly those regularities that scientists classify as laws as opposed to those they regard as accidental? These sound very much like broadly empirical claims about scientific practice—very much the sort of descriptive/interpretive investigation that I told you that I was interested in when our conversation started. So maybe we are not so far apart after all. The claims about metaphysics and grounds/truth-makers that motivate the BSA are at least in part tied to empirical claims about the content and practices of the various sciences, especially physics.

M: Well, sort of, but I have to admit things are a bit more complicated. It turns out for example, that when you apply the BSA to cosmology, various regularities (such as those having to do with large scale isotropy and uniformity of mass-distribution of the universe) that most cosmologists seem to regard as accidental or non-lawful turn out to be laws after all (cf. Callender 2004). So if the BSA is correct, we should conclude that these scientists were wrong in classifying these regularities as non-laws.

JW: I can certainly understand why naturalistically minded philosophers of science prefer the BSA to inflationary metaphysical programs that populate the universe with relations of necessitation between universals and their ilk. I applaud the attempt to provide an account of laws that connects with features of scientific practice. But I don't want to accept the BSA simply on the grounds that it appears less metaphysically extravagant than its competitors. And in this connection, what you've just said concerns me a bit. I mean first you motivate the account by appealing to certain standards that you claim guide scientific practice in the identification of laws, and then when those standards apparently lead to generalizations being identified as laws that are not so regarded in scientific practice, you conclude that the practice is wrong. Why don't you instead conclude that your account of the practice is wrong—maybe laws don't have to do with trade-offs of simplicity and strength in the way that you claim.

M: While it is true that, as I emphasized earlier, many of us like our work in metaphysics to be science-inspired, we certainly don't want to be uncritical slaves to the pronouncements of scientists. Maybe cosmologists are just wrong in regarding generalizations about the large-scale isotropy (and so on) of the universe as non-laws. That is what our best theory of laws of nature suggests. You don't think we should automatically defer to everything scientists say, do you? As heirs to a proud humanistic tradition that stretches back thousands of years, we should be self-confident enough to avoid such "scientism."

JW: It would help me to understand how the BSA is supposed to work if we could work through a simple example.

M: I'm happy to oblige. Consider a record of the positions, velocities, and masses of the sun and all of the various planets in the solar system throughout all time. The Humean Basis will include this sort of information. We want to systematize as much of this information as possible within a framework that achieves a best balance of simplicity and strength. Now this record will include the information that the planets (if we include the asteroid Ceres) exhibit a regular pattern or spacing in the distance of their orbits from the sun—this is what is known as Bode's law.[10] Bode's law thus describes a regularity. Consider adding this to the Best Systemization as an axiom. The result of doing so would be (let us assume) a gain in strength since we will now be able to use this regularity to deduce more facts about the behavior of the planets. At the same time, addition of Bode's law to the BS would result in a loss of simplicity since the systemization now includes an additional axiom. In assessing this trade-off, we "see" that this loss in simplicity outweighs the gain in strength, hence that the Bode's "law" does not belong to the Best Systemization and is consequently not a law.

JW: This is ingenious but I'm not convinced that it is a very accurate description of the way in which the epistemic values that guide the discovery of laws relate to one another. For one thing, even if we put aside the fact that we do not have a good account of simplicity of the sort this framework requires, theory construction in science does not seem to be guided by the kind of trade-off between simplicity and strength you describe. Rather than a trade-off, science (and particularly fundamental science) often seems guided by an ordering of these two considerations in which strength has something close to lexical priority over simplicity. Other things being equal, it is permissible, even required, to add complexity, even a lot of complexity, when this contributes to even "small" (assuming we know how to measure this) increases in strength. Consider Newton's first rule of reasoning: "We are to admit no more causes of natural things than such as are both true, and sufficient to explain the appearances" (Newton 1726/1729, 202). This rule seems to say that if some additional complexity (in the form of an additional cause) is required to explain appearances, we should admit it, but not otherwise. The rule does not say that we should *omit* to introduce additional causes that are sufficient to explain appearances if the simplicity gain resulting from this omission outweighs the resulting loss in strength, which is what the trade-off postulated in the BSA seems to require. Instead, Newton seems to be suggesting that

[10] Here I am merely reporting a story that is sometimes used to motivate the BSA. In fact, the story is inaccurate, since Bode's law does not correctly predict the orbit of Neptune.

the introduction of additional causes/complexity is always justified (perhaps required) if this leads to an improvement in strength. Your picture of trading off simplicity and strength seems to license arguments like the following: my theory doesn't predict very much about planetary positions or movements, but it is so much simpler than theories that do that it achieves a better overall balance of simplicity and strength and is preferable for that reason. Isn't it obvious that this is not supported by generally accepted ideas about good scientific methodology?

Moreover, if this is right, then your story about Bode's law does not work. Instead, on the assumption that strength has (near) lexical priority over simplicity, it seems we should conclude that Bode's law and other similar generalizations should after all be admitted as axioms in the Best Systemization (since their addition to the BSA would lead to an increase in strength), in which case they will count as laws, contrary to what everyone seems to agree is the correct assessment of their status.[11]

M: I agree that the details of the BSA may require a little tweaking—some of us are working on that. The important thing to bear in mind, though, is that simplicity and strength are important in theory-choice and that, whatever the details may be, we can use them to provide a properly reductive account of laws. I'd advise you to focus on these larger points and not get too distracted by niggling details. The BSA is the most metaphysically promising approach to laws of nature that we have, particularly since it is appropriately reductive. So it *must* be right, at least in broad outline. Otherwise we would have a situation in which the laws of nature lack metaphysical foundations and are ungrounded, which is obviously intolerable.

JW: Let me try a slightly different angle. You've said that the BSA is to be thought of as a systemization of the entire Humean Basis—everything that happens in the entire universe over its entire lifetime.

M: That's right.

JW: But obviously scientists don't have access to any thing like this when they formulate and test theories. So in this respect at least, the BSA does not describe how science actually proceeds.

M: It's a thought experiment. We project from what the small corner of the universe to which we presently have access to everywhere and everywhen and then think in terms of competing systemizations of that.

---

[11] For more detail, see Woodward 2014.

JW: But this Humean Basis we end up imagining—it just involves at bottom what you describe as point-wise arrangements of fundamental physical magnitudes or something like that? And this is understood in such a way that no characterizations making reference to any kind of physical modality is allowed into the characterization of the basis?

M: That's right. If we let that stuff in, we would lose the possibility of a reduction. We would be caught in a circle.

JW: But now it looks as though it is your aspiration to produce a reduction rather than anything grounded in how scientists actually reason about laws that is driving your characterization of this supposed basis. In one respect, the HB includes far more information (about what happens everywhere in the universe over its entire lifetime) than scientists in fact have access to. In another respect, the Basis seems to include far less information, in comparison with the evidence and other considerations on which scientists actually rely in reasoning to conclusions about laws, causal relationships, or other claims with modal content. What I mean is that when I look at the ways in which scientists in various disciplines, including physics, infer or reason to conclusions with modal content, they often seem to make use of some version of the following schema:

> Prior assumptions about modal (causal or nomological) relationships *p* + non-modal information *I* about regularities, correlations, initial and boundary conditions → conclusions about other modal relationships (new modal knowledge) *p**.

For example, prior causal knowledge (e.g., that *Y* does not cause *X* and that *X* and *Y* are not caused by some third variable *Z*) can be combined with information about correlations (e.g., that *X* and *Y* are correlated) to infer that *X* causes *Y*. Information about the trajectories of the planets of the sort represented by Kepler's laws can be combined with very general nomologically committed assumptions (e.g., the motion of the planets is caused by some central force due to the sun) to derive the gravitational inverse square law, as Newton himself showed.

If your characterization of the HB excludes all such prior causal or nomological information, how can it capture how scientists actually reason when they make causal or nomological inferences? And in general if our aim is to capture "our" (or scientists') notion of law, shouldn't we expect that notion to reflect or be shaped by the information that we actually have access to in reasoning about laws, rather than information we do not have access to?

M: I don't deny that scientists often reason on the basis of the considerations you describe. However, it is obvious that it must be possible in principle to reconstruct what they are doing so that it fits the BSA. For example, if we take one of these assumptions—call it p—with modal content that you are talking about and test it by making use of another assumption with modal content q and so on, then the worry is that we either (i) are led to a regress (testing assumption q requires modally committed assumption r, testing r requires modal assumption s, and so on) that never ends up being grounded just in what is non-modal, or else (ii) we end up with a circle (having to assume p again). Neither possibility is satisfactory from the point of view of justification. The only remaining alternative is that the modal assumptions on which we rely must ultimately be justified in some other way, and this can only involve some sort of balancing of simplicity and strength. Thus the inferences you describe ultimately must be reconstructable within the BS framework.

JW: I don't understand why you are so confident that it must be possible to carry out this reconstruction.

M: If it is not possible to do so, you will be relying on modal claims that are not reducible, and reasoning involving such claims will be unclear, involved in a regress, and circular. Science would be Ungrounded.

JW: First, I'm puzzled about why you phrase things in terms of whether it is *possible* to provide such a reconstruction or grounding or whatever, rather than in terms actually producing the reconstruction. If there is worry about modal claims being unclear and so on, why don't we need to actually exhibit the details of the reconstructions/reductions you describe in order to assuage these worries? Or at least provide enough of the details to make it plausible that such reconstructions are always going to be possible? But you've just agreed that the scientists employing modal claims do not in fact actually ground them in the way you describe and apparently don't feel any need to produce reductions of the modal to the non-modal. Moreover, you haven't yet pointed to any epistemic or conceptual calamities that ensue from these omissions. Indeed, if there were such calamities, why wouldn't they appear in scientific practice and lead scientists themselves to be concerned to address them, which I take it that we both agree does not happen? In the absence of such defects, why isn't the obvious moral that the sort of reduction you are talking about just is not required if what one is interested in is elucidating how scientists and others provide evidence for, test, and reason about causal and nomological claims?

I can see why, given the way that metaphysicians conceive of what they are doing, there is a motivation within metaphysics for requiring that grounds/

truth makers and perhaps reductions be produced, but it sounded earlier as though you were claiming something more than that—that such truth-makers need to be provided if scientific inquiry is to avoid circles and regresses, or to "make sense" or avoid fatal unclarities or be adequately grounded at some deeper level or something like that.[12] Is your view perhaps that doing metaphysics but not science requires providing grounds and truth-makers? I'm happy to agree with that, as long as you agree that I am not required to do metaphysics

I'll add that, given that, as I think you have to acknowledge, the notions of "simplicity" and "best balance" (and, for that matter, "strength") that figure in the BSA are not exactly pellucid, I don't see that a reduction of modal notions via the BSA, even if it could be provided, would somehow endow these notions with a "clarity" and "intellectual respectability" that they would otherwise lack, at least in any ordinary sense of those words. I think perhaps you are confusing "sufficiently clear and unambiguous to guide reasoning, at least for certain purposes and contexts," with the desirability of providing a certain kind of reduction.

M: This is getting tedious. As I have been trying to explain, the cogency and legitimacy of the demand for reductive truth-makers for modal claims is acknowledged by many of the greatest philosophers, past and present. It is central to metaphysics. Only an ontological philistine would resist this demand.

JW: OK. Maybe I could get a better handle on what the BSA involves if we could discuss how it bears on another issue that I'm interested in. This has to do with the status of the various generalizations that figure in the so-called special sciences. These generalizations seem to have limited domains of application and typically have exceptions even within the domains in which they are intended to apply. Nonetheless we make distinctions among these generalizations—some seem to describe causal relations and are accorded a central role in theory construction and causal explanation, while others, although no more exception-ridden, are not regarded as having a different status. For example, it is uncontroversial that changes in the money supply are usually correlated with changes in the inflation rate—there is a true or approximately true generalization describing this correlation. But in the comparatively recent past there was considerable disagreement

[12] Might we think of Metafisico's view as something like the following: scientists do not provide reductions of the modal notions they use and typically have no opinions about (and don't seem to care about) whether it is possible to do so. Nonetheless, (i) it *is* possible to do so, and (ii) it is only because this is possible that the use of (unreduced) modal notions in science is legitimate. As nearly as I can see, there is nothing incoherent about this position. Nonetheless, it offers (and I see) no reason to believe either (i) or (ii). Thanks to Matthew Slater for some helpful comments here.

among economists about whether changes in the money supply cause changes in the inflation rate or whether instead the causality runs in the opposite direction from inflation to money or whether instead the correlation between these two factors is due to some third factor. What does the BSA tell us about the difference (if any) between the claim that money and prices are correlated and the claim that money causes prices? If it is true that money causes prices, would it be appropriate to think of this as a *law* of economics, as some philosophers of social science have suggested, or should we be thinking about it in some other way? To take another example, if I was an economist contemplating the Phillips curve describing the inverse relationship between inflation and unemployment in the period before 1970, would the BSA tell me if I should regard this generalization as an economic law? Or whether we should think of it as describing a causal relationship rather than a mere correlation?

M: The status of generalizations in the special science is an interesting question and one that is relatively unexplored within the BSA, which has focused mainly on laws in fundamental sciences like physics. Still, several suggestions have recently emerged and are the subject of ongoing discussion. It is probably most useful to begin with the version of the BSA formulated by Lewis (e.g., 1999) and others. On this version, it seems doubtful whether there are any laws in the special sciences. For one thing, recall that the BSA characterizes laws in terms of a notion of simplicity that is tied to the perfectly natural properties; the laws will be generalizations that are simple when framed in terms of such properties. Presumably the perfectly natural properties are, at least to a large extent, those that figure in fundamental physics. It is unlikely that any of the properties that figure in the generalizations of the special sciences are perfectly natural. When formulated in terms of perfectly natural properties from physics, the generalizations of the special sciences are likely to be horrendously complicated and non-simple. On this ground alone, they are disqualified as laws. To this it may be added that, often if not always, whether these special science generalizations hold seems to depend on whether various physically contingent initial and boundary conditions hold. These initial and boundary conditions are likely to be very non-simple and to contribute relatively little to strength, since in many cases they will hold only very locally—e.g., in the case of the psychology and economics, perhaps only for human beings and a few other organisms. So those initial/boundary conditions are unlikely to make it into the Best System. Hence quite apart from the point about non-natural properties, the generalizations of the special sciences are unlikely to be derivable as theorems of the best system and hence are not laws.

JW: I see. I'm afraid, though, that this doesn't seem to help with the problems I described. As I said, there does seem to be distinction between those true generalizations in the special sciences that describe causal relationships (and, at least according to me,[13] have other features like stability or invariance under changes) and those true generalizations that do not have these features. These seem like distinctions that are tracking something in the world. That is, it looks to me as though there is a difference between, on the one hand, the way the world is if it is true that money causes prices and, on the other hand, the way that it is if money does not cause prices. And there are important issues about the sort of evidence that would be relevant to establishing which if either of these claims is true—that is why the social and behavioral sciences are full of procedures for inferring and testing causal claims. Even if it is correct that there are few laws in the special sciences—a claim that I have defended myself—if the application of the BSA to the special sciences classifies all true generalizations in the special sciences into an undifferentiated group as "non-laws," it does not seem to help with the issues that interest me. Put differently, on this version of the BSA, the law/non-law distinction does not seem to correspond to or to illuminate the causal/merely correlational distinction that I would like to understand.

M: The version of the BSA we've been discussing may not help with the problems you describe, but it does have a consequence that is far more metaphysically important: if correct, it shows that there are no laws outside of physics. This is a claim about what the world is like on a fundamental level and thus of the greatest ontological significance. Surely you don't claim that the question of whether there are any laws outside of physics is unimportant?

JW: I don't think that asking whether the generalizations found in disciplines outside of physics are "laws" is the most useful way of formulating issues about of the nature and status of those generalizations. In general, I think that you metaphysicians focus far too much on the notion of law. I do think there are facts about whether various relationships studied in the special sciences are stable in various ways, about whether these can be exploited for purposes of manipulation and control, and it is these features that philosophers of science should be interested in. I don't much care about whether you call these "laws." Refusing to call these relationships "laws" doesn't make them disappear or make them unworthy of study. On the other hand, merely bestowing the honorific "law" on them doesn't do much to elucidate them either.

---

[13] Cf. Woodward 2003.

M: Whatever. If you don't like the consequences for the special sciences of the version of BSA just sketched, here is a variant on this account that leads to results that you may find more appealing. It is called the Better Best Systems account (Cohen and Callender 2009), and in effect it relativizes the BSA to each of the special sciences.[14] Take some special science of interest—for example, economics or ecology with its proprietary predicates and its associated Humean Basis described in terms of these predicates. Consider alternative systemizations $\{S_E\}$ of this Basis. The laws of the science in question are then just those axioms (or axioms and theorems) that are common to the systemizations in $\{S_E\}$ that achieve a best balance of simplicity and strength. In fact, we can think of this framework as supplying the underlying metaphysics that is missing from your talk of invariance and relations exploitable for purposes of manipulation and control. Moreover, the respectability of the special sciences is restored, since we have shown (or it least we have shown how it might be possible, which is all that matters in metaphysics) that they contain laws. This result should please you, as someone who is interested in these areas of inquiry and seems to take them more seriously than they probably deserve.

JW: I don't mean to sound ungrateful, but I still have the feeling that the problems that interest me still haven't been addressed. If I have understood correctly, you would like to conceptualize the question of whether money M causes prices P as a claim about whether there is a law linking M and P. This law claim will be true or not depending on whether this generalization is an axiom or theorem in the BBS. If instead, some appropriate generalization linking P to M was such an axiom or theorem, then it would be the case that P causes M. Unlike the BSA, the BBS thus at least allows for the possibility that one of these claims might turn out to be a law.

M: That's right.

JW: I still don't see how to *use* the BBS (or the BSA) to determine which of these alternative law claims is correct, especially since the notions of simplicity and best balance on which you are relying have been left so vague and underspecified.

---

[14] In the interest of fairness and completeness (and to avoid leaving a misleading impression about his views), I should point out that Callender has been quite critical of some aspects of work in contemporary analytical metaphysics. See his (2011). It is also true, as he has emphasized to me in correspondence, that many advocates of the BSA present it as an anti-metaphysical view of laws or at least an approach that carries minimal metaphysical commitments with it. I agree that it is a virtue of the BSA that it avoids the kind of metaphysics that postulates non-naturalistic special entities and relationships. On the other hand, at least from my perspective, the BSA retains some important metaphysical commitments, including its aspirations to provide a reduction of the modal to the actual and its reliance on ideas about "perfectly natural" predicates. It remains more metaphysical than some of its advocates realize.

Maybe it would help if I posed matters this way: Econometricians make use of various sorts of tests ("causality tests") and appeal to certain kinds of evidence to try to determine whether M causes P or vice-versa. What does the BBS have to say about these? It just does not seem very helpful in clarifying what the disagreement in this case is about or what sort of evidence is relevant to settling it. Does the application of the BBS/BSA to the special sciences show which of these "causality tests" are good ones for identifying causal relationships? Can we actually use it to distinguish genuine causal relationships from mere correlations in social science contexts?

M: It may well be true that the BBS can't be used to determine which of these competing causal claims is correct if that is what you want to know. But *if* it is true that, for example, money causes prices, the BBS can be used to capture or represent the metaphysical status of this generalization in a satisfactory way, rather than leaving it obscure and mysterious, ontologically speaking, which is what you have done.[15] More generally, I certainly wouldn't expect metaphysics to be "useful" in the sense you seem to have in mind or for metaphysical treatments of causation to cast light on issues of evidence and testing. Perhaps you might explain more clearly what it is that you are looking for. I've always thought that a sharp separation between, on the one hand, what is merely pragmatically useful (or of merely epistemic or methodological significance, as considerations having to do with evidence and testing are) from the point of view of the parochial interests of human beings from, on the other hand, issues

---

[15] I owe to an anonymous referee an alternative suggestion about how Metafisico might reply, which appeals to a Lewisian account of causation:

> M : Here is another way of thinking about your question: If a change in the money supply caused a change in the inflation rate, then there would be a counterfactual relationship between them: roughly, if the money supply had not changed, the inflation rate would not have changed. Of course, it's a little more complicated than that, but that's the basic idea. In this respect, my view isn't so very different from your interventionist view, JW. Except for two things. First, I think that I can spell out the details without any danger of circularity—I won't have to mention causation when articulating the details of this counterfactual relationship. Secondly, I have a view about what grounds the counterfactual truth—that's a matter of closeness relations among possible worlds. Now, if you had any gift at all for metaphysics, JW, you would be asking me what grounds these closeness relations. I can tell you in detail if you want, but a key feature of close-by possible worlds is that they share the same laws or at least very similar laws. And what grounds the laws? That's where the BSA comes in, which means that we can ground everything we've been talking in what we call the Humean mosaic. So even if all these layers of grounding are not of interest to you (though they should be to any philosopher), the first bit of grounding, the grounding of causes in counterfactuals, may well help with your problems and interests.

In the interest of space, my reply will be abbreviated. First, the criteria for closeness of worlds employed in this Lewis-type analysis (perfect match of matters of fact, match of laws, etc., and the trade-offs among these) are far too vague and unclear to guide inference to causal relationships in connection with real-life cases like the relationship between money and prices (Woodward 2016). Relatedly, the

about what is metaphysically true is one of the central achievements of contemporary analytic metaphysics. The Sublime Uselessness of our subject has always been one of its most appealing characteristics for me.

JW: Here is an illustration of what I have in mind in talking about "usefulness." As I mentioned earlier, the interventionist approach to causation that I favor associates causal claims with claims about what the outcomes of hypothetical experiments would be, were an intervention to occur on the putative cause variable. I see this as having a number of implications for the methodology of constructing and testing causal claims. Let me focus on the latter—testing was not the main focus of my book, but it may help to suggest what I am driving at when I claim that an interventionist account of causation of the sort that I favor can be methodologically useful and when I worry that the sort of metaphysical foundations you are looking for are not likely to be useful for these sorts of purposes. Suppose that we are interested in using non-experimental evidence *E*, in conjunction with other information *I*, to assess whether *X* causes *Y* is true. Then the question we should ask ourselves, according to interventionists, is whether *E* and *I* provide reason to think that, if a properly controlled experiment were to be performed in which *X* is manipulated, *Y* would change. In other words, in cases of causal inference from non-experimental data, we should think of ourselves as trying to infer from that data what the results of an experiment would be without actually carrying out the experiment—we use the hypothetical experiment to clarify what would be required to establish the causal conclusion on the basis of non-experimental information. In fact, there are formal results about this: it is possible to show that, given certain natural assumptions, we can infer that *Y* will change under an intervention on *X* from appropriate non-experimental data about correlations, given certain background assumptions. For example, we can infer to this conclusion if there is some third variable *Z* which is (i) known to cause *X*, (ii) known to cause *Y* if at all only through *X*, and (iii) which is such that it is uncorrelated with other causes of *Y* (except for those variables if any on the causal route from *Z* to *X* to *Y*). A variable meeting these conditions is called an *instrumental variable* by econometricians, and if *Z* is such an instrument,

"laws" associated with the money-prices relationship, which would need to figure in the assessments of closeness for purpose of applying the counterfactual analysis, are completely unknown. Second, to the extent that assessments of closeness are clear, the Lewisian analysis gives the wrong results in particular cases (Woodward 2003). Roughly, this is a consequence of what Metafisico supposes to be an advantage of the Lewisian analysis—its reductive character. It appears that you can't characterize a properly behaved closeness metric without appealing to causal information.

and *X* and *Y* are correlated, we can infer that *X* causes *Y*. Indeed, we can provide a quantitative estimate of the "treatment effect" of *X* on *Y* from information about the covariances of *X*, *Y* and *Z*: effect of *X* on *Y* = *Cov (Y,Z)/Cov (X,Z)* (cf. Winship and Morgan 2007).

The relevance of this to interventionist treatments of causation is that the underlying logic is readily understandable in interventionist terms : one can show that under conditions (i)–(iii), *Z* satisfies (modulo a complication[16]) my conditions for an intervention on *X* with respect to *Y*; thus (according to the interventionist characterization of causation) if *X* and *Y* covary under an intervention on *X*, then *X* causes *Y*.

There are many, many other examples of how thinking of problems of causal inference in non-experimental contexts as though one were trying to ascertain what the results of a hypothetical experiment would be without actually doing the experiment can help to elucidate what sort of evidence is required for reliable causal inference and can disambiguate and clarify competing causal claims

I'll add that I don't mean to single out the BSA or BBS in suggesting that they don't seem very relevant to epistemological or methodological issues associated with causal claims and laws. Saying that the truth makers of causal claims (or laws) are facts about dispositions or relations of necessitation between universals seems equally unilluminating for purposes of understanding how we can use evidence to assess causal claims, use such claims reliably in reasoning, and so on.

M: Once again, I think you misunderstand the goals that we metaphysicians have in looking for truth-makers and foundations. We are not trying to address issues of non-experimental or experimental design or to evaluate statistical techniques for testing causal claims. We are looking for something much deeper—an understanding of what is going on at the level of fundamental metaphysics.

JW: Fair enough—I get that you are not trying to do methodology and also that you think that metaphysics of science should be done in a way that

[16] The complication is that although it follows from conditions (i)–(iii) that conditions (I1), (I3) and (I4) in my characterization of an intervention (Woodward 2003, 98) are satisfied, it does not follow from (i)–(iii) that my remaining condition (I2) (that the intervention act as a switch or arrow-breaker with respect to X) is satisfied; indeed, in many real-life cases instrumental variables will not be switch-like. Fortunately, Eberhardt (2007) in effect has shown that one may replace (I2) with a weaker condition (characterizing a so-called soft intervention), which does not require that interventions be switch-like, but only that they supply an exogenous source of variation in the variable intervened on that is uncorrelated with the other causes of that variable. This notion of a soft intervention enables many of the same inferences as one can make in the presence of arrow-breaking interventions and also can be used to define various causal notions characterized in Woodward 2003. Instrumental variables are very often interventions in this weaker sense.

abstracts from all merely methodological or practical concerns. What I do *not* get is why you also insist that I have to provide an account of the grounds or metaphysical foundations for causal claims before I can address the epistemological and methodological issues that interest me. On the contrary, given our discussion so far, it looks to me as though even if we had the right story about the grounds or truth makers for causal claims, this would not help with the issues that interest me.

M: Metaphysics does indeed bear an asymmetric relationship to the other topics you mention, but this should not be surprising in view of its foundational role. Briefly, epistemology and methodology are not at all relevant to metaphysics, but metaphysics is highly relevant to them, insofar as they have any philosophical interest. However, the details of how this (the bearing of metaphysics on method) works does not matter for present purposes and, in any case, is not the concern of the metaphysician, as one can see by examining virtually any contemporary discussion of the metaphysics of causation. But rather than continuing in this vein, let me try another tack. I take it from what you said above about interventions that a necessary condition for I to count as an intervention on X with respect to Y is that intervention I causes X.

JW: That's right.

M: But then your account, which characterizes causation in terms of the response of Y to interventions on X, is blatantly circular.

JW: I acknowledged in *MTH* that my account is non-reductive, although I also pointed out that the causal information that figures in the characterization offered of what it is for X to cause Y has to do with other relationships besides the $X \rightarrow Y$ relationship. In other words, the characterization suggests how you can use information about other causal relationships to assess whether X causes Y, but does not require that you have to already know whether X causes Y to use the characterization. Thus the circularity involved in my characterization, such as it is, is not vicious, at least for epistemic purposes—indeed, at the level of particular causal claims, there is no circle at all, although there is also no reduction of those claims to claims that are non-causal.[17] Incidentally, the example involving instrumental variables above provides a concrete illustration

[17] The topic of circular definitions deserves much more attention that I can give it here. As Anil Gupta has long argued (e.g., 2015), there are any number of "definitions" in mathematics and science that are formally "circular" but nonetheless provide useful information and illumination. I owe to Mark Wilson the following example: i = the unique number x such that $x^2 = -1$ & x = i (personal communication); this is "circular," yet it captures the core behavior of the complex numbers. In general, even if "circular," a definition can function so as to exclude some cases from falling under the defined term and include others.

of this point—use of this method (when understood as above) requires prior causal knowledge, but it can be used to provide evidence for different causal relationships. So a non-reductive account can be quite useful for epistemological or methodological purposes. Again my conjecture, based on our conversation so far, is that accounts of laws and causal relationships that are reductive in aspiration are particularly unlikely to be helpful in clarifying the methodological and epistemological issues that interest me, both because the notions that figure in reductive accounts (simplicity and so on) are so unclear and because in the interests of making the reduction work philosophers seem to end up misrepresenting what it is they are trying to reduce, as with the idea that good scientific method involves a trade-off of simplicity and strength. I think that a similar point holds for the non-reductive accounts of truth makers that you have mentioned.

M: I'm afraid that you completely lack an ear for metaphysics. You are one of the most ontologically shallow people I have ever met. (Aside: It is not vouchsafed to just anyone to discern the Lineaments of Being.)

JW: I fully agree. Let me buy you a drink.

ACKNOWLEDGMENTS

I am grateful to Bob Batterman, Jim Bogen, Craig Callender, Chis Hitchcock, Peter Machamer, John Norton, Matthew Slater, Kyle Stanford, Martin Thomson-Jones, and Mark Wilson for comments on an earlier draft. In this case, the usual disclaimer that they do not necessarily endorse the ideas expressed in this dialogue applies with extra force.

REFERENCES

Armstrong, D. 1983. *What Is a Law of Nature?* Cambridge: Cambridge University Press.

Callender, C. 2004. "Measures, Explanation and the Past: Should 'Special' Initial Conditions Be Explained?" *British Journal for the Philosophy of Science* 55: 195–217.

Callender, C. 2011. "Philosophy of Science and Metaphysics." In *The Continuum Companion to Philosophy of Science*, edited by S. French and J. Saatsi, 33–54. London: Continuum International Publishing Group.

Chalmers, D., D. Manley, and R. Wasserman. 2009. *Metametaphysics: New Essays on the Foundations of Ontology*. Oxford: Oxford University Press.

Cohen, J., and C. Callender. 2009. "A Better Best System Account of Lawhood." *Philosophical Studies* 145 (1): 1–34.

Correia, F., and B. Schnieder. 2012. *Metaphysical Grounding: Understanding the Structure of Reality*. New York: Cambridge University Press.

Dretske, F. 1977. "Laws of Nature." *Philosophy of Science* 44: 248–68.

Eberhardt, F. 2007. "Causation and Intervention." PhD diss., Carnegie Mellon University. Available at philosophy.artsci.wustl.edu/people/Frederick_Eberhardt.

Goldenfeld, N., and L. Kadanoff. 1999. "Simple Lessons from Complexity." *Science* 284: 87–89.

Gupta, A. 2015. "Definitions." In *Stanford Encyclopedia of Philosophy*. Stanford University, 1997. Article published April 10, 2008, revised April 20, 2015. http://plato.stanford.edu/entries/definitions/.

Heil, J. 2003. *From an Ontological Point of View*. Oxford: Oxford University Press.

Holland, P. 1986. "Statistics and Causal Inference." *Journal of the American Statistical Association* 81: 945–60.

Kendler, K. 2005. '"A Gene for...': The Nature of Gene Action in Psychiatric Disorders." *American Journal of Psychiatry* 162: 1243–52.

King, G., R. Keohane, and S. Verba. 1994. *Designing Social Inquiry*. Princeton, NJ: Princeton University Press.

Lewis, D. 1999. *Papers in Metaphysics and Epistemology*. Cambridge: Cambridge University Press.

Lowe, E. 2011. "The Rationality of Metaphysics." *Synthese* 178: 99–109.

Manley, D. 2009. "Introduction: A Guided Tour of Metametaphysics." In *Metametaphysics: New Essays on the Foundations of Ontology*, edited by D. Chalmers, D. Manley, and R. Wasserman, 1–37. Oxford: Oxford University Press.

Morgan, S. and C. Winship. 2007. *Counterfactuals and Causal Inference: Methods and Principles for Social Research*. Cambridge: Cambridge University Press.

Mumford, S. 2004. *Laws in Nature*. London: Routledge.

Newton, I. 1726. *Philosophiæ Naturalis Principia Mathematica*, 3rd ed. Andrew Motte, English translation, 1729.

Pearl, J. 2009. *Causality: Models, Reasoning, and Inference*. Cambridge: Cambridge University Press.

Rubin, D. 1974. "Estimating Causal Effects of Treatments in Randomized and Non-Randomized Studies." *Journal of Educational Psychology* 66: 688–701.

Schaffer, J. 2009. "On What Grounds What." In *Metametaphysics: New Essays on the Foundations of Ontology*, edited by D. Chalmers, D. Manley, and R. Wasserman, 347–83. Oxford: Oxford University Press.

Spirtes, P., C. Glymour, and R. Scheines. 2000. *Causation, Prediction and Search*, 2nd ed. Cambridge, MA: MIT Press.

Sider, T. 2009. "Ontological Realism." In *Metametaphysics: New Essays on the Foundations of Ontology*, edited by D. Chalmers, D. Manley, and R. Wasserman, 384–423. Oxford: Oxford University Press.

Strevens, M. 2007. Review of *Making Things Happen*, by J. Woodward. *Philosophy and Phenomenological Research* 74 (1): 233–49.

Tooley, M. 1987. *Causation: A Realist Approach*. Oxford: Oxford University Press.

Van Inwagen, P. 2007. "Metaphysics." *Stanford Encyclopedia of Philosophy*. http://plato.stanford.edu/archives/win2016/entries/metaphysics/.

Winship, C. and Morgan, S. 2007. *Counterfactuals and Causal Inference*. New York: Cambridge University Press.

Woodward. J. 2003. *Making Things Happen: A Theory of Causal Explanation.* New York: Oxford University Press.

Woodward, J. 2014. "Simplicity in the Best Systems Account of Laws of Nature." *British Journal for the Philosophy of Science* 65: 91–123.

Woodward, J. 2015. "Methodology, Ontology, and Interventionism." *Synthese* 192: 3577–99.

Woodward, J. 2016. "Causation in Science." In *Oxford Handbook of the Philosophy of Science*, edited by P. Humphreys, 163–84. New York: Oxford University Press. Longer online version: DOI:10.1093/oxfordhb/9780199368815.013.8.

10

Against Bracketing and Complacency

METAPHYSICS AND THE METHODOLOGY OF THE SCIENCES

Martin Thomson-Jones

1. SOME QUESTIONS, AND THE BRACKETING STRATEGY

There are at least two ways metaphysical questions can arise in philosophical thinking about the sciences. First, we can ask metaphysical questions when we ponder the interpretation of particular scientific theories: Are species individuals in evolutionary theory? Do many-particle systems violate the principle of the identity of indiscernibles in quantum mechanics? Is time an emergent phenomenon according to string theory or loop quantum gravity? Second, we can ask metaphysical questions about notions which show up in connection with many parts of the sciences, whether in scientific talk or in philosophers' accounts of the sciences at large. Much of the work which has been going on under the heading "metaphysics of science" in recent years deals in questions of the second sort. For example, there has been quite a bit of work on questions centering on such notions as *law of nature, causation*, and *disposition*. Among the central questions have been these: Are there such things as laws of nature? Causal relations? Dispositions? If so, what kinds of things are they? And how are they connected, to the extent that they are?[1]

The questions I want to consider here arise in the second way—they concern notions which show up in scientific talk in many parts of the sciences, and in philosophers' accounts of the sciences at large. Attempting to answer these

[1] Of course, the questions themselves have a much longer history.

questions, which focus on such notions as *model, mathematical structure*, and *fictional object*, has not generally been seen as part of an investigation of the metaphysics of science, and they have been less widely pursued than the questions about laws, causation, and dispositions. These questions are recognizably metaphysical nonetheless, and they parallel the more familiar questions perfectly: Are there such things as models, mathematical structures, and fictional objects? If so, what sorts of things are they? And how are they connected, to the extent that they are?[2]

At the same time, there are differences between questions about dispositions, laws, and causation, on the one hand, and the questions about models, mathematical structures, and fictional objects, on the other. Questions in the first cluster are at least sometimes questions about the *content* of our scientific pictures of the world. Some think that the sciences, or many of them, picture the world as containing or being governed by laws; others disagree. Among those who agree, there is disagreement about what sorts of things laws must be, and how they are related to causation and to dispositions, given the way the sciences picture the world. The same goes for causation and for dispositions, *mutatis mutandis*. Here, then, the "metaphysics of science" is the metaphysics of the scientific world-picture, or of the various scientific world-pictures.

The questions I want to consider here, however, might be described as questions about the metaphysics of science as an activity. Models, mathematical structures, and fictional objects are not part of the content of our scientific pictures of the world; rather, if they exist, they are among the tools we use to paint the pictures. Scientists, in talking about the things they do, describe themselves as constructing models, and as employing both mathematical structures and fictional, imaginary, or hypothetical objects to model and to represent.[3] And when philosophers of science offer their own, usually more systematic accounts of scientific representation, modeling, explanation, confirmation, and various other scientific activities, some of them say the same sorts of things. It is now a commonplace in the philosophy of science that scientists construct and use models; some philosophers of science present accounts of scientific representation or of modeling which place a central emphasis on the role of mathematical structures; some offer accounts of at least some important varieties of scientific representation on which those kinds of representation involve the use of fictional

[2] Here and throughout, and purely for the sake of concision, I will use the term "model" in an artificially restricted sense, to narrow attention to *nonconcrete* models. That is, I am not addressing metaphysical questions about the existence or nature of those physical objects which we class as models, such as orreries, or Crick and Watson's tinplate model of DNA.

[3] Hereafter, and again for the sake of concision, I will use the term "fictional object" to cover talk of fictional, imaginary, and hypothetical objects.

objects.[4] Presented with scientific talk and philosophical accounts of these sorts, we can ask whether there are such things as models, mathematical structures, and fictional objects; if so, what sorts of things they are; and how they are related. In asking such questions, we can again be said to be investigating the metaphysics of science, but this time we are investigating the metaphysics of scientific practice.[5]

These last questions, then, are the questions which provide my starting point. My concern here, however, is not to try and answer these questions about the metaphysics of scientific practice; I and others have made attempts in that direction elsewhere.[6] Instead, I want to consider a particular reaction to the raising of these questions. Some philosophers of science are prone to brute expressions of impatience or disgust when they get a whiff of metaphysics, but that is not the reaction I have in mind. Rather, I want to consider the claim that, whatever the merits or demerits of such questions in themselves, there is simply no need, from the point of view of the philosophy of science, to grapple with them. On this view, we can give satisfying answers to the methodological and epistemological questions which concern us in the philosophy of science—questions about scientific representation, modeling, explanation, confirmation, and so on—without answering questions about the existence and nature of models, mathematical structures, or fictional objects, *even if* our answers to the philosophy-of-science questions appeal to such entities.[7] The proposal, then, is that we employ what I will call the *bracketing strategy*: the strategy of putting aside such questions as irrelevant to our purposes in the philosophy of science.[8] I want

---

[4] See, for example, van Fraassen (2008) and Weisberg (2013) on mathematical structures, and Godfrey-Smith (2006), Thomson-Jones (2007) and (forthcoming), Contessa (2010), and Frigg (2010) on fictional objects. (There is a question about the precise sense in which Frigg's and Godfrey-Smith's accounts picture some varieties of scientific representation as involving fictional objects, but that is a question for another time; all that matters here is that both authors talk at central points as though some varieties of scientific representation involve the use of fictional objects.)

[5] Note, however, that the questions about laws, causation, and dispositions can also be motivated by methodological interests. Consider, for example, the idea that making sense of confirmation or of explanation involves reliance on a substantive distinction between laws and accidental generalizations, or accounts of experimental testing which draw on particular views about causation (such as the account in Cartwright 1989). This is why I wrote, in the preceding paragraph, that such questions are "at least sometimes" questions about the content of our scientific pictures of the world.

[6] See Thomson-Jones (1997), (2007), (2010), and (2012) on what sorts of things models can be, and their relations to mathematical structures, and Thomson-Jones (2007) and (forthcoming) on models and fictional objects. Contessa (2010), Frigg (2010), and Toon (2012) also address questions about the existence and nature fictional objects, and their relationship to models, and Weisberg (2013) discusses the relationship between models, mathematical structures, and fictional objects.

[7] Putting things this way implies that the metaphysical questions about the existence and nature of models, mathematical structures, and fictional objects are not questions for the philosophy of science. My aim here is to challenge that claim, in one sense at least (see section 2). But it is part of the view I am characterizing here.

[8] A weaker version of this proposal insists that we need not answer questions about the existence and nature of models, mathematical structures, or fictional objects in order to provide satisfactory accounts of, say,

to examine a particular justification for this strategy, one which rests on the claim that any adequate account of ordinary and scientific discourse will legitimize our ordinary and scientific ways of talking about models, mathematical structures, or fictional objects—call this the *legitimization argument*.[9] In consequence, the idea goes, we philosophers of science can simply get on with constructing accounts of scientific representation, modeling, theory structure, idealization, confirmation, explanation, and the rest, and rely on talk about such objects when it is useful to do so, safe in the assumption that when philosophers of language, metaphysicians, philosophers of mathematics, and aestheticians settle on satisfactory accounts of such talk, we will simply be able to add those accounts as appendices to our own work. The questions we are trying to answer in the philosophy of science are thus seen as swinging free of the metaphysical issues in question—call this the *autonomy thesis*.

This sort of maneuver is familiar throughout philosophy, of course, and the questions which get bracketed can be of any kind—that is, they need not be metaphysical. Perhaps because the move is so standard, invocations of it with respect to talk of models, mathematical structures, and fictional objects in the philosophy of science typically only arise in conversation, when the metaphysical questions are pressed; or at least, that has been my experience. But one can also find explicit recommendations along the lines just presented. So, for example, Bas van Fraassen advocates the adoption of such a strategy by philosophers of science with regard to worries about the existence and nature of mathematical objects, and he offers exactly the sort of justification I have described, insisting that "*for any philosophy of mathematics to be acceptable it must imply that the ordinary use of mathematics is fine*" (2005, 97). Thus we can talk about functions and vector spaces and set-theoretical *n*-tuples in offering accounts of theory structure, representation, and the like, and leave it to philosophers of mathematics to decide the answers to a range of further questions about the interpretation of such talk.[10] Similarly, Peter Godfrey-Smith, in pursuing the idea that we might think of model systems as "imaginary objects" on par with fictional characters, seems happy to engage in our ordinary ways of talking about

---

representation, modeling, confirmation, explanation, and the like, but allows that we might need to answer such questions for certain other purposes we have, or might have, in the philosophy of science. (And there are certainly disagreements about the proper purposes of philosophy of science out there.) My arguments will be arguments against either version.

[9] Note the (inclusive) disjunction, "models, mathematical structures, or fictional objects." Someone who offers an account of scientific representation which involves talk of mathematical structures but not of fictional objects, for example, will be able to justify her bracketing of metaphysical questions about the latter quite straightforwardly. See section 3 for a more careful laying-out of the legitimization argument; and thanks to Kyle Stanford for catching a slip on this point.

[10] Anjan Chakravartty has also suggested to me that we should take this sort of approach to mathematical discourse in philosophical accounts of scientific representation (pers. comm.).

fictional characters while developing an account of scientific modeling, and to defer the task of settling on an acceptable interpretation of that talk until later, when we may wish to take it on for "general philosophical reasons" (2006, 734–36; quoted phrases from 735).[11]

There may well be situations in which employing this sort of strategy is unproblematic, but I want to argue that this is not one of them.

2. THE PRACTICAL NECESSITIES OF PHILOSOPHICAL LIFE

Before I present my arguments against the legitimization argument, the autonomy thesis, and the bracketing strategy, however, I should make it clear what I am objecting to, and what I am not.

A bit of terminology: for the sake of concision, I will use the term "core accounts" for accounts of scientific representation, modeling, theory structure, idealization, explanation, confirmation, and the like—all aspects of scientific practice which are among the central epistemological and methodological focal points of the philosophy of science. This choice of label is not meant to imply that constructing accounts of such things is *the* core of the philosophy of science; it is intended only to reflect the fact that the project of constructing accounts of such things is at the center of some large part of the discipline, and closer to its core (these days, at least) than the attempt to construct accounts of the semantics and ontology of our ordinary and scientific discourse—the other sort of account to which I will need to make frequent reference.

---

[11] Godfrey-Smith is not explicit about taking bracketing to be justified by the assumption that any acceptable account of our talk about fictional entities will legitimize the sorts of things we ordinarily say, but it is hard to see how the approach he recommends could be a reasonable one without that assumption. No such legitimization assumption would be required, of course, if Godfrey-Smith were aiming only to characterize a particular way of thinking and talking that scientists engage in; but he is also aiming at an understanding of what is happening when scientists think and talk in those ways, and he is recommending an account which itself engages in talk of "imagined concrete things" (2006, 734–35).

Incidentally, Godfrey-Smith does say that "[a]t the end of the day . . . some general account must be given of the imagined objects of both ordinary fiction and scientific modeling" (2006, 735), but given that he mentions ordinary fiction and scientific modeling in the same breath here, and given that he is only requiring that a "general" account be given "[a]t the end of the day," I still read him as endorsing the idea that we can address our central concerns as philosophers of science without such an account, even if (to put the point in my terms) we engage in talk of fictional objects in the process.

Note also that Godfrey-Smith's comments appear in just the sort of context I have in mind: one of his immediate philosophical concerns is to offer an account of a certain representational strategy we find employed in the sciences—that is, to offer at least a partial account of how representation works in science. Another, closely related to the first, is to say something about the nature of models. Similarly, van Fraassen is defending his use of mathematical ways of talking in the philosophy of science (and other parts of his philosophical work) in the passage from which the remark I have just quoted is drawn.

First, then, I am not saying that all work on modeling, say, should grind to a halt until certain metaphysical questions about the existence and nature of models, mathematical structures, and fictional objects have been answered. It is a common enough situation in philosophy, as in other fields of inquiry, that the answer to one question depends in part on the answer to a number of other, equally difficult and uncertain questions. And it is a perfectly sensible way of coping with such situations, as a temporary measure, simply to assume something about the answers to those others questions. (Exactly what one assumes will not usually be an arbitrary matter, of course; typically it is a sort of educated bet.) Accordingly, I have no quarrel here with the philosopher of science who says, for example, "Here's my account of scientific representation/modeling/theory structure/explanation/ . . . ; note that it presupposes a certain sort of nominalism about mathematics," or "note that it presupposes that there are fictional objects, and that they are abstract artifacts."[12] Of course, this does mean that in order to fully evaluate the account on offer, we will have to take a stand on whether the presupposed account of mathematical or fictive and metafictive discourse is correct; but again, that is a common enough situation in philosophy, and I have no axe to grind in this respect.

Someone who employs what I am calling the bracketing strategy is doing something quite different, however. Rather than making certain metaphysical commitments without necessarily regarding it as decisively established that they are the right commitments to make, she presents a core account which engages quite heavily in talk of models, mathematical structures, or fictional objects, and yet she simultaneously denies that she has thereby taken on substantive commitments with respect to the ontology of modeling discourse, mathematical discourse, or fictive and metafictive discourse. She thus insists that we can provide core accounts which provide satisfying answers to the questions we have as philosophers of science, and which do so in part by making free use of talk of models, mathematical structures, or fictional objects, without having to address the metaphysical issues in question. This is the claim that is my target here. And my aim in this paper is to critique a justification that has been offered for this autonomy thesis and the corresponding bracketing strategy, and, moreover, to make a case that the thesis is false and the strategy illegitimate.

A second, smaller point of clarification: neither am I insisting that philosophers of science who present the sorts of accounts in question should become metaphysicians (or philosophers of mathematics, philosophers of language, or aestheticians), even part-time. One reasonable way of proceeding is to turn to other philosophers who have thought long and hard about the nature and existence of mathematical

---

[12] I have in mind, with the latter example, Amie Thomasson's account of the ontology of fiction; see her *Fiction and Metaphysics* (1999) and my (forthcoming).

structures, fictional objects, and the like for guidance on the answers to the metaphysical questions. That is one way of dealing with the fact that, if we as philosophers of science offer core accounts which rely heavily on talk of models, mathematical structures, or fictional objects, we have become hostage to the outcome of certain investigations in metaphysics. The most I claim is that we must admit that fact, and be responsive to it in some way or other.

These points of clarification out of the way, let me now turn to the central business of critiquing the legitimization argument and, furthermore, making it plain why I think the autonomy thesis false, and the bracketing strategy unjustifiable.

3. THE LEGITIMIZATION ARGUMENT

It will help to begin by laying out the legitimization argument more explicitly. Some new terminology will smooth the way: the *as-if practice for X's* is the practice of talking as if there are *X*'s, and as if they have certain sorts of features, or stand in certain sorts of relations (to one another or to other things). I will have in mind throughout cases in which the *X*'s are models, mathematical structures, fictional objects, or some combination of these (though I suspect my arguments would generalize to other kinds of putative entity). For example, Alma, a philosopher offering an account of scientific representation, might say,

> Physicists often use mathematical structures to represent the systems they are investigating via relations of isomorphism; for example, they will use a function from an interval on the real line to vectors in an infinite-dimensional Hilbert space to represent the evolution of the state of an electron over time.

In uttering this sentence, Alma is talking as if there are mathematical structures, and as if they can stand in relations of isomorphism to physical systems; she is thus engaging in the as-if practice for mathematical structures. Similarly, Lester, a philosopher offering an account of scientific modeling might say,

> One way scientists model real pendula involves exploring the ways in which they are similar to and differ from such fictional objects as the simple pendulum.

Lester is talking as if there are fictional objects, and as if they can stand in relations of similarity to physical systems; he is thus engaging in the as-if practice for fictional objects.[13] One important note: the label "as-if practice" should not be taken to imply

[13] See n. 4 for some philosophers who write in these sorts of ways.

that the ways of talking to which it applies are mistaken or misleading; it is meant to leave that question open.

We can now set out the legitimization argument. Consider Iris, a philosopher of science who is offering an account of some aspect of scientific practice—scientific representation, modeling, explanation, confirmation.... Suppose that in offering her account, Iris is engaging in the as-if practice for *X*'s (models, mathematical structures, or fictional objects). Iris insists, however, that she can bracket questions about the existence and nature of *X*'s for the purposes of her work as a philosopher of science. To support this insistence, she offers the legitimization argument:

(1) Any adequate account of ordinary and scientific discourse will legitimize our ordinary and scientific ways of talking about *X*'s.
(2) When I engage in the as-if practice for *X*'s in offering my account of this aspect of scientific practice, I am engaging only in ordinary and scientific ways of talking about *X*'s. So, from (1) and (2):
(3) Any adequate account of ordinary and scientific discourse will legitimize my use of the as-if practice for *X*'s in offering my core account. So, from (3):
(4) *Autonomy thesis for X's*: I don't need to answer questions about the existence and nature of *X*'s in order to provide a satisfying account of the aspect of scientific practice with which I'm concerned. So, from (4):
(5) For the purposes of developing a satisfying account of this aspect of scientific practice, I can adopt the *bracketing strategy for X's*: I can put aside questions about the existence and nature of *X*'s as questions I don't need to address (even if, in developing my account, I engage in the as-if practice for *X*'s).

This seems to be the sort of justification many philosophers of science have in mind when they claim that they can set aside questions about the existence and nature of models, mathematical structures, or fictional objects even when considering core accounts in which talk of such putative entities plays a central role.

Before presenting my objections to this argument, I should address two questions about claims (1) and (3): First, what does it mean to call something an "adequate account of ordinary and scientific discourse" in this context? And second, what is it for an account of ordinary and scientific discourse to "legitimize" our ordinary and scientific ways of talking about *X*'s? It is worth noting that any difficulties that might be involved in providing full answers to these perfectly good questions are, after a certain point, grist for my mill—they would simply provide additional grounds for doubt about the argument, over and above the grounds I will identify in the next section. Still, my own arguments in what follows will assume that the relevant notions are to be taken in a certain sort of way, at least, and so it would be a good idea to be explicit about what I am assuming along those lines before moving on.

First, then, I am assuming that an account of our ordinary and scientific discourse will answer various semantic questions about the utterances involved in that discourse. On the assumption that we engage in the as-if practice for *X*'s as part of our ordinary and scientific discourse, this will mean answering some central semantic questions about the utterances we produce when engaging in the as-if practice for *X*'s—our *as-if utterances about X's*, as we might call them. And that will involve answering some questions about the existence and nature of *X*'s.

An *ideal* account of our ordinary and scientific discourse, I take it, will then have all the following features:[14] it will make sense of the ways we ascribe truth values to the utterances making up the discourse (when we do), and of patterns in any other, less direct ways we have of treating such utterances as true, or false, or lacking a truth value;[15] it will make sense of our epistemic attitudes to the utterances making up the discourse (making sense, for example, of the fact that we regard some utterances as knowable and others as unknowable, some as known and others as unknown);[16] it will be internally consistent; it will employ an acceptable ontology, or at least not employ an unacceptable one;[17] and it will both display internal coherence, and cohere with our best accounts of other kinds of discourse, at least where it seems it should.

Adequacy, of course, is a lower bar. Perhaps some of the features just listed are features that even a merely adequate account must have (internal consistency, for example), and perhaps adequacy is then a matter of having a sufficiently large proportion of the other features, or having enough of them to a sufficiently great degree (making sense of *most* of our ascriptions of truth values to utterances, for example). But saying this much is, I think, enough for the purposes of the arguments that follow.

What is it for an account of our ordinary and scientific discourse to legitimize our ordinary and scientific ways of talking about *X*'s (step [1]) or, more specifically, our engagement in the as-if practice for *X*'s (step [3])? I will focus on just the latter half

---

[14] (i) Perhaps there are other features an ideal account of our ordinary and scientific discourse should have, too; we need not settle that here. (ii) For the sake of brevity, I will write as though utterances themselves can have truth values, and can properly be treated as knowable or unknowable, known or unknown, and so on; what follows could easily be rephrased to accommodate the idea that it is rather the propositions expressed by our utterances which have such features.

[15] This is not to say that an ideal account will deliver the judgment that our ascriptions of truth-values are correct, or that we are right to treat utterances as true, false, or lacking a truth value in the particular ways that we do; it might entail that we are mistaken in some or all cases, and perhaps go on to explain why we do not realize our mistakes.

[16] Which, similarly, is not to say that an ideal account will deliver the judgment that our epistemic attitudes are correct.

[17] After all, what we find acceptable ontologically in the end might in part be a matter of seeing what the ontological commitments of an otherwise-adequate account of ordinary and scientific discourse are.

of this question, as the as-if practice is the part of our ordinary and scientific way of talking about *X*'s that matters here. Now the most obvious way of at least making a start on legitimizing our engagement in the as-if practice would be to show that our as-if utterances are true (or at least those we seem to produce assertorically).[18] But the proponent of the legitimization argument surely does not want to begin by assuming that any adequate account of our ordinary and scientific discourse will do *that*; surely she means to leave open the possibility that our as-if utterances about models, mathematical structures, or fictional objects are useful fictions, say. Otherwise, premise (1) of the argument is a highly controversial claim (and, one might add, a claim it might be hard to support without abandoning the bracketing strategy, which the argument is intended to justify). So I will take it that the proponent of the legitimization argument means to allow that legitimization could take any one of a number of forms. More specifically, I will take it that an account of our ordinary and scientific discourse could legitimize our ordinary and scientific ways of talking about *X*'s in part by showing that our as-if utterances about *X*'s, or a sufficiently large proportion of them

- are true, or
- are useful fictions, in the sense of being useful falsehoods,[19] or
- though unknowable, are useful,[20] or
- though strictly meaningless, are apt, in the sense that they obey the rules of the (useful) game we are playing when we produce them,[21] or
- fall into one of the preceding categories.[22]

---

[18] I want to allow this might only amount to making a start on the task of legitimizing our engagement in the as-if practice for two reasons. First, to talk about showing that our as-if utterances are true is implicitly to limit attention to *declarative* as-if utterances. This is the right place to put the emphasis in the present context, because declarative utterances take center stage when we construct our core accounts; but legitimizing the as-if practice for *X*'s overall might involve dealing with other kinds of as-if utterance—questions, commands, and so on. Second, full legitimization of the as-if practice might involve treating things other than individual utterances: e.g., it might involve showing that the inferences we perform when engaging in our ordinary and scientific ways of talking about *X*'s (or a sufficiently large proportion of them, anyway) are good, in some sense of "good."

[19] Cf., for example, the view that (virtually) all our ethical utterances are false, but valuable nonetheless; or, rather closer to home, the parallel view about our mathematical utterances.

[20] I have in mind a view on which our as-if utterances are taken to be truth-valued, but on which we cannot know their truth values. It might nonetheless be thought useful in one way or another to engage in the practice of producing utterances: in the service of producing an empirically adequate and explanatorily powerful account of some aspect of scientific practice, say (perhaps given an appropriately extended notion of empirical adequacy, and noting that it will be controversial whether explanatory power is compatible with unknowability). Here van Fraassen's constructive empiricism—a view about scientific theorizing about the world—is serving as a model of a view one might take about certain sorts of philosophical theorizing about science.

[21] Cf. old-time semantic instrumentalism about theoretical utterances in the sciences.

[22] That is, the legitimizing account could show that all our as-if utterances fall into the same one of the preceding categories, but without establishing which; or it could merely establish the weaker claim that any given as-if utterance falls into one of the preceding categories.

This list of varieties of legitimization may well not be exhaustive; in what follows, I will assume only that everything on it is indeed a variety of (partial) legitimization.

Why, even on this more inclusive notion of legitimization, should we believe premise (1)? The best way of making this premise seem plausible, I suspect, is as follows: "It is clear that engaging in our ordinary and scientific ways of talking about *X*'s is (at least) a useful practice, and one with some reasonably well-defined internal rules. Given that this is clear, any account of our ordinary and scientific discourse, in order to be deemed adequate, will have to capture that fact—namely, the fact that talking in our ordinary and scientific ways about *X*'s is a well-defined and useful practice—and so will have to legitimize those ways of talking in at least that sense."[23] This line of support might strike us as more persuasive for some values of *X* than others: I take it that it is clearer in the case of mathematical structures that our ordinary and scientific talk about *X*'s is a well-defined and useful practice than it is in the case of models or fictional objects. In all three cases, however, this way of supporting premise (1) seems promising enough that, at least for the purposes of this discussion, I will put aside any further doubts of this sort.[24]

We might also ask about the support for premise (2). What is required, note, is that in any particular case in which (i) a core account is offered which engages in the as-if practice for *X*'s, and (ii) the legitimization argument is invoked to justify the bracketing of questions about the existence and nature of *X*'s nonetheless, the core account in question must engage in the as-if practice for *X*'s only in ways which are part of our ordinary and scientific ways of talking about *X*'s. Whether this requirement is satisfied in this or that particular case will be a moot issue, however, if my main line of criticism succeeds.

4. AGAINST THE LEGITIMIZATION ARGUMENT, AND AGAINST BRACKETING

I now want to argue, then, that even if we grant its premises, the legitimization argument fails—neither the autonomy thesis for *X*'s (step 4) nor the claim that we can

---

[23] Perhaps this amounts to legitimization of the fifth sort just listed.

[24] Except this one: Faced with this way of supporting premise (1), one might wonder why we should be confident that there will *be* an adequate account of our ordinary and scientific discourse which captures the (putative) fact that talking in our ordinary and scientific ways about *X*'s is a well-defined and useful practice. After all, in searching for an adequate account we are at least hoping to find an account which employs an acceptable ontology, and which coheres with our accounts of other kinds of discourse. This makes adequacy sensitive to our ontological beliefs, and our beliefs about other kinds of discourse; and some of those beliefs may be wrong, even assuming that they we have not adopted them without justification.

The response to this worry, I take it, is to say that if there is no account of our ordinary and scientific discourse which both captures the (putative) fact that our ordinary and scientific ways of talking about *X*'s is a well-defined and useful practice *and* coheres with our ontological beliefs, or our beliefs about other kinds of discourse, then given the (putative) self-evidence of the (putative) fact, we have good reason to reject the

engage in the as-if practice for *X*'s and bracket with impunity (step 5) follows. The first of my two main conclusions will thus be that we lack a justification for the unobvious claim that we can bracket questions about the existence and nature of *X*'s when putting forward core accounts which engage in the as-if practice for *X*'s. Along the way, an argument will emerge that that claim is, moreover, false, and so my second and stronger conclusion will be that the bracketing strategy is in fact unavailable to us in such situations.

Suppose, then, that we know that premise (1) is true: we know that any adequate account of our ordinary and scientific discourse will legitimize our ordinary and scientific ways of talking about *X*'s. Suppose we also know that, as premise (2) maintains, the as-if practice for *X*'s is encompassed by our ordinary and scientific ways of talking about *X*'s.[25] It follows that any adequate account of our ordinary and scientific discourse will legitimize the use of the as-if practice for *X*'s in our work as philosophers of science, constructing accounts of scientific representation, modeling, explanation, confirmation, or the like (step [3]). And to make things as propitious as possible for the bracketer, suppose we know, too, that there is at least one adequate account of our ordinary and scientific discourse, and that the correct account is an adequate one, so that we have compelling reason to believe that the as-if practice for *X*'s is in fact legitimate.[26] The idea we are examining is then that if we find ourselves in such an epistemic situation, it follows that (i) we can engage in the as-if practice for *X*'s in constructing our core accounts, and furthermore, that (ii) in doing so, we need not address questions about the existence and nature of *X*'s, and (so) need not adopt any particular account of the semantics or ontology of that practice.[27] I will argue, however, that neither thing follows.[28]

My arguments will rest, first, on the observation that there is more than one variety of legitimization, and second, on the assumption that in constructing accounts of scientific representation, modeling, explanation, and so on, we are aiming to know the truth about those things, and to understand them. (This assumption

---

"coherence" constraints on adequacy (because we then have good reason to doubt that our ontological beliefs, or our accounts of other kinds of discourse, are correct).

[25] Strictly speaking, the assumption we need here is just that the as-if practice for *X*'s *as it is employed by the particular core account at hand* is encompassed by our ordinary and scientific ways of talking about *X*'s; different accounts might engage in different versions of an as-if practice about *X*'s. I will leave this qualification aside to simplify the discussion.

[26] For a reason that these additional assumptions are not entirely trivial, see n. 24.

[27] I am taking it here that adopting an account of the semantics and ontology of the as-if practice for *X*'s would in part involve addressing questions about the existence and nature of *X*'s.

[28] The conjunction of claims (i) and (ii) here may not be exactly equivalent to the conjunction of claims (4) and (5) in the formulation of the legitimization argument presented in the preceding section, but both (i) and (ii) follow from the conjunction of (4) and (5) (at least when taken in context), and so we can show that (4) and (5) do not follow from the premises of the argument by showing that (i) and (ii) do not.

about the aims of our philosophical work may not meet with universal agreement, but I will not attempt to defend it here.) The problem is, first, that if one or more of the adequate accounts of ordinary and scientific discourse which (we are supposing) we know there to be legitimizes the as-if practice in the wrong way, (i) may still be false, for without knowing more we cannot be sure that engaging in the as-if practice will not be at odds with aiming to know the truth; and second, that even if all adequate accounts legitimize the practice in the right way, so that engaging in the as-if practice is perfectly compatible with our interest in knowing the truth, (ii) will be false nonetheless, as engaging in the as-if practice without further insight into the semantics and ontology of that practice will make it impossible to achieve the aim of understanding scientific representation, modeling, explanation, and so on.

The crucial distinction among varieties of legitimization here is between those which rule out the knowable truth of as-if utterances about *X*'s and those which do not. So, for example, some adequate account of ordinary and scientific discourse might legitimize the as-if practice by making it clear that, although speaking as if there are models, mathematical structures, or fictional objects is *ipso facto* a matter of uttering falsehoods, say, or of uttering meaningless sentences, it is useful nonetheless to treat some such utterances as true in the right context.[29] If we normally take some of the as-if utterances in question to be true, an account which legitimizes the as-if practice in one of these ways might even come equipped with a story about why we are so easily gulled into doing so—an error theory. Regardless, legitimization of either of these two sorts is incompatible with the supposition that our as-if utterances are true, and so, *a fortiori*, precludes the possibility that they are knowably true.

The same goes for an account which legitimizes by insisting that utterances about models, mathematical structures, and fictional objects make claims which, though truth-valued, are unknowable, but then goes on to explain how the claims in question can be "acceptable" in something like the sense in which some scientific claims about the unobservable world are acceptable according to the constructive empiricist—a sense of "acceptable" in which P's acceptability does not entail its truth, its falsehood, or its believability.[30] An account of our ordinary and scientific

[29] See the second and fourth entries on the list of varieties of legitimization given in the preceding section.

[30] See, for example, van Fraassen 1980, 12–13. Note that van Fraassen speaks there only of acceptance, rather than of acceptability; by calling the claim P "acceptable," I mean just that acceptance in van Fraassen's sense is a reasonable attitude toward P. And I say "something like" the constructive empiricist's sense of "acceptable" in order to leave open the question of whether the notion of empirical adequacy is directly applicable in the present context, or whether some other kind of adequacy might here play the role that empirical adequacy plays when it is claims about the unobservable that are up for assessment.

discourse which legitimizes the as-if practice in this way, too, precludes the possibility that our as-if utterances are true and known to be.[31]

Now, if an account of ordinary and scientific discourse legitimizes the as-if practice in a way which involves ruling out the knowable truth of the claims we make when we engage in that practice, then given that account, claim (i) will be false: we cannot engage in the as-if practice in constructing accounts of scientific representation, modeling, explanation, and the rest, given that we are aiming to know the truth about the targets of those accounts in constructing them, for the accounts will then necessarily contain either false claims or unknowable ones. If, for example, we offer an account of scientific representation which commits us to uttering sentences like "Scientists use fictional objects to represent real systems in the domain of inquiry," but utterances which involve speaking as though there are fictional objects are all false, or unknowable, then at least some important part of our account of scientific representation is either false or unknowable.[32]

---

31 This notion of legitimization is suggested not only by van Fraassen's constructive empiricist account of scientific discourse about unobservable objects and events, but also by the view that the best way to make sense of mathematical discourse might be to see it as aimed at something other than "true story-telling," a view van Fraassen describes himself as "sympathetic to" in "On Taking Stances: An Interview with Bas van Fraassen," the Harvard Review of Philosophy piece cited above (2005, 96–97). The notion of legitimization just set out clearly goes beyond the latter sort of view, however: saying that the aim of a kind of discourse is something other than true story-telling is compatible with saying that the utterances which make it up uniformly express false or truth-valueless claims, or fail to express claims at all (and indeed, it is even compatible with saying that they express knowable truths).

32 Note that in this case it would be misleading to say that simply that we do not have a justification for the bracketing strategy, as though an appropriate response would be to continue to look for a justification, or to drop the bracketing strategy in favor of investigating the semantics and ontology of the as-if practice right away. Rather, in this case we ought simply to drop the as-if practice itself, for the purposes of the philosophical tasks at hand. (This is not to say that we would have reason to eschew the as-if practice in every context—perhaps it would be fine to go on engaging in it at the right moments in our scientific work, for example.)

As I noted earlier, some may reject the metaphilosophical assumption that we are aiming to know the truth in constructing philosophical accounts of the nature of models, or theory structure, or idealization. (Rosen [1994, esp. 151–52] lays out a reading of van Fraassen on which he is not aiming at the truth in presenting constructive empiricism; see van Fraassen [1994] for his reaction.) I am not sure I know how to defend that assumption. One might appeal to the distinct idea that we are aiming to understand, say, scientific representation by constructing an account of it, and then try to argue that an account cannot yield understanding if it rests on claims which are false, or unknowable, so that aiming to know the truth is part of aiming at understanding. But this is to take on contentious issues, too. van Fraassen, for example, clearly thinks that unknowable claims can explain, as he thinks that claims about the unobservable can explain, but are unknowable (1980, chap. 5); it would not be surprising, then, if he were to insist that unknowable claims can produce understanding. And Frisch (1998) argues that the best explanations are sometimes false. (Of course, Cartwright has argued that explanation often involves false law statements, too [e.g., 1983, essay 2], but as I read her, she has gone on to reframe her view. As she conceives it in her later work, the point is just that if the law statements we appeal to in many of our explanations are read in the traditional way, as purportedly exceptionless generalizations about what happens, then they are false. We should, however, read them differently, as telling us about the capacities that various properties carry with them, and so read, law statements face no special obstacle to truth.)

Consequently, if, for all we know, one or more of the adequate accounts of ordinary and scientific discourse which (we are supposing) we know there to be legitimizes the as-if practice in a way which precludes the knowable truth of the as-if utterances, and we know no more, then for all we know the *correct* account of the as-if practice is one which rules out knowable truth.[33] And this means that, for all we know, engaging in the as-if practice in developing accounts of representation, modeling, explanation, and the rest is entirely at odds with aiming to know the truth about those things. We will thus not be justified in continuing to engage in the practice in such an epistemic situation.

Suppose instead, then, that we do know, somehow, that all adequate accounts of ordinary and scientific discourse legitimize the as-if practice for *X*'s by reassuring us that a good range of the utterances we produce when engaging in that practice make claims which are both true and knowable.[34] Being in *this* epistemic situation does not give us reason to eschew the as-if practice in constructing our core accounts; but that is not to say that we will then be justified in employing the bracketing strategy. Establishing that engaging in the as-if practice is not at odds with achieving our philosophical aims does not amount to showing that we can achieve those aims without an articulated understanding of the as-if practice. And I want to argue that if we engage in the as-if practice when developing our core accounts, then those core accounts will yield an understanding of their targets (scientific representation, modeling, explanation, etc.) only when supplemented by an understanding of the as-if practice. Bracketing is thus at odds with the immediate aim of understanding representation, modeling, explanation, and the rest.[35]

---

33 The clause "and we know no more" rules out cases in which we somehow know that the *correct* account of our ordinary and scientific discourse allows for (or even guarantees) the knowable truth of the as-if utterances. It is hard to see how we would know this without having done enough thinking about the semantics and ontology of the as-if practice to have already abandoned the bracketing strategy.

34 Here I have moved from considering cases in which an account of ordinary and scientific discourse precludes the knowable truth of the as-if utterances to considering cases in which knowable truth is entailed. There is a third possibility, of course, at least logically speaking: an account might legitimize the relevant discourse, in some recognizable sense of "legitimize," while leaving it open whether the as-if utterances are knowably true. At this point, however, it is easy to see that it would be impossible to justify the bracketing strategy if one or more adequate accounts of our ordinary and scientific discourse were, for all we knew, of this third sort. If one or more of the legitimizing accounts leaves open the knowable truth of the claims we make when engaging in the as-if practice, and we bracket further investigation, then for all we know, everything we say when we engage in the as-if practice is false, or unknowable. And as long as we are ignorant on that score, we cannot come to understand or know the truth about modeling, scientific representation, explanation, or anything else by engaging in the as-if practice.

35 That is, although given this sort of legitimization there is no special problem with part (i) of the idea we are examining (that we can engage in the as-if practice in constructing our core accounts), I want to argue that there is still a problem with part (ii) (that in doing so, we need not adopt any particular account of the semantics or ontology of that practice).

To see this, consider the two metaphysical questions which lie at the center of an inquiry into the semantics and ontology of modeling discourse, mathematical discourse, or fictive and metafictive discourse: Are there objects which are the referents of the noun phrases of that discourse ("the simple pendulum model," "the Hilbert space which is the state space of the proton," "Emma Bovary")? And if so, what sorts of things are they? Call these *the existence question* and *the nature question*, respectively. The issue at hand is whether we can arrive at an understanding of scientific representation, explanation, and so on by speaking as though there are such objects and as though they have certain features, knowing (somehow) that we are speaking truly in doing so, but putting off the task of answering those two questions.

Take the existence question first. To show that this question cannot be bracketed consistently with our philosophical aims, I will take as an example an account of modeling which employs talk of fictional objects;[36] the argument carries over, *mutatis mutandis*, to any account of modeling which centrally relies on talk about models, mathematical structures, or fictional objects, and to accounts of idealization, theory structure, the nature of models, explanation, confirmation, and so on which do so.

Suppose, then, that we have an account of modeling which centers on claims expressed by sentences such as this:

> (S) Scientists often model real pendula by exploring the ways in which they are similar to and differ from fictional objects such as the simple pendulum.

In having us utter sentences such as (S), the account engages in the as-if practice for fictional objects. That fact alone does not preclude our knowing the account to be true, given the assumption we are now making that a good range of the utterances which make up the as-if practice express knowably true claims. The account might be false nonetheless, of course (or, perhaps, true but unknowable), as the assumption in question does not guarantee the knowable truth of the claims expressed by any particular set of utterances; but in that case appealing to it would be at odds with the aim of knowing the truth about modeling. So let us suppose then, that the account is knowably true; suppose, in fact, that we somehow know it to be true. The remaining problem is simply that without an answer to the existence question, we cannot claim to know what this account of modeling *says*, and so we cannot claim to have arrived an understanding of modeling by considering it.

---

[36] Cf. Godfrey-Smith (2006) and Frigg (2010). Both authors tend to prefer the terms "imaginary object" and "imagined object," but my talk of fictional objects is meant to include imaginary objects, and in any case, both authors introduce the talk of imagined objects as part of a project of drawing an analogy between scientific modeling and ordinary fiction.

The argument for this last claim is as follows. Either there are fictional objects such as the simple pendulum, or there are not. If there are, then presumably sentence (S) should be taken literally, along with the rest of the account of which it is a part. If there are not, on the other hand, then scientific modeling does not involve such objects; and as the sentence in question says otherwise on its surface, then (given our assumption that the sentence is making a true claim) the sentence is not to be taken literally, and nor, by extension, is the account of which it is a part. So if we do not know whether there are fictional objects—if we do not know the answer to the existence question—we cannot know whether the account of modeling is to be taken literally. We also do not know how the account should be taken if it is not to be taken literally. If there are no fictional objects, and so no such fictional object as the simple pendulum, but (S) is true nonetheless, then it is not *obvious* what (S) means; so what does it mean? We surely cannot claim to have arrived at an understanding of modeling by invoking such an account in the midst of such fundamental uncertainty about how the account should be taken. Yet removing that uncertainty will at least involve answering the existence question about fictional objects, and so will require us to reject the bracketing strategy.[37]

What about the nature question? Suppose we somehow managed to convince ourselves that there are fictional objects, thus settling the existence question; could we at least then bracket the question of what sorts of things fictional objects are? Again, the answer is no—not if we hope to come to understand modeling by relying on an account which engages in the as-if practice for fictional objects by centering on claims expressed by sentences such as (S). Inserting different stories about the nature of fictional objects into such an account will yield different final pictures—not only of the sort of thing to which scientists compare real systems when modeling, but of what the comparing involves, and of the kinds of similarity and difference relations that can obtain between fictional objects and real systems. And different pictures of these things make for different ways of understanding of how modeling works.

[37] The point can be put a little more rigorously if we are willing to grant a certain assumption about meaning and linguistic understanding, as follows: If there are fictional objects such as the simple pendulum, then the simple pendulum presumably features in the truth conditions of (S), which will have to do with that fictional object's standing in certain relations to scientists and real pendula. If there are no such things as fictional objects, on the other hand, then clearly no fictional simple pendulum features in the truth conditions of the sentence. Thus if we do not know whether there are such things as fictional objects, we cannot know whether the truth conditions of sentence (S) involve a thing of that sort; and so we cannot know the truth conditions of the sentence. Given the right assumptions about meaning and linguistic understanding, it follows from this that we do not really know what (S) means, and that we do not understand it. Consequently we do not understand the account of modeling we have been offered, and so cannot come to understand modeling itself by considering that account. Thus bracketing of the existence question is not an option if we are to pursue such an account.

Three cases will suffice to make the point. (a) Extrapolating from Lewis (1978) and Lewis (1986), we might take the view that fictional objects are concrete inhabitants of other concrete possible worlds.[38] In that case, the simple pendulum can stand in straightforward relations of similarity to actual pendula by sharing certain properties with them, or by having "nearby" determinate values of such shared determinables as period of oscillation. (b) According to Thomasson (1999), fictional objects are abstract artifacts. On that approach, the simple pendulum, though it exists, does not have a mass or a period of oscillation, and so cannot stand in straightforward relations of similarity to concrete pendula. Instead, modeling concrete pendula will involve comparing the properties of those concrete pendula with the properties the simple pendulum has *according to the relevant fiction* (properties it does not in fact *have*).[39] (c) Given the emphasis on talk of "imagined" objects in Godfrey-Smith (2006) and Frigg (2010), we might consider the view that fictional objects are mental objects of some sort. In that case, too, the simple pendulum will have neither a mass nor a period of oscillation, and so the full story about how it can stand (or be said to stand) in similarity relations to concrete pendula will again have to be more complex than the surface simplicity of an utterance like (S) suggests.

Different answers to the nature question about fictional objects will thus lead to different pictures of modeling, even given that the starting point is the same in each case—namely, a set of utterances like (S). And this example makes it doubtful, I think, that we could claim to have achieved the aim of understanding modeling by appealing to any account which relies on speaking as though there are fictional objects until we have provided an answer to the question of what sort of things fictional objects are (unless, of course, we have already answered the existence question negatively). The same goes, I take it, for other as-if practices, and for accounts of scientific representation, idealization, explanation, confirmation, and other targets of our interest in the philosophy of science. Once again, then, we see that bracketing is at odds with our immediate philosophical aims in constructing core accounts.

One objection to this line of argument focuses on the nature question, and runs as follows: "Certainly there are differences between the pictures of modeling which result from (S) by supplying different accounts of the nature of fictional objects; but it is not clear that those differences will make a difference to anything we care about as philosophers of science. In each of the three cases, (a)–(c), we get a picture on

---

[38] I should stress the word "extrapolating" here, as this is not something Lewis himself claims. On the other hand, he does propose that the idealized systems spoken of in scientific theorizing be taken to be such concrete possibilia (1986, 26–27). See Thomson-Jones (2007, §3.6) for a fuller discussion.

[39] Thomasson (2003) offers a variant of her (1999) view on which the story would be different again. See also Thomson-Jones, forthcoming.

which fictional objects can stand in similarity relations of *some* sort to real objects, and on which scientists can *in some way or other* compare fictional objects with real; and perhaps that is all we need to know about this aspect of modeling in order to understand the things we are interested in as philosophers of science. So it is not clear that we cannot bracket the nature question in such a case."

I am tempted to reply simply by saying that one of the things we are interested in as philosophers of science is how modeling works, and as we get different pictures of *that* by supplying an (S)-type account with different accounts of the nature of fictional objects, the differences among the pictures do make a difference to at least one of the things we are interested in as philosophers of science. I suspect, however, that this response would do little to address the concerns of someone who takes the objection to have some force in the first place. This might be because such a philosopher finds little interest in the project of understanding how modeling works per se, and is interested in understanding modeling only insofar as doing so will help us to answer various other methodological and epistemological questions. For the sake of argument, then, suppose one sides with the philosopher presenting the objection on the matter of what questions or topics are of ultimate interest in the philosophy of science; still, there is more to be said.

It is first worth pointing out exactly what the objection establishes if it succeeds. At most, it shows that it *may* be that we can bracket the nature question in the situation at hand; for all this objection says, it may instead turn out that we cannot. That will hinge on whether the differences between the pictures of modeling we get by supplying (S)-type accounts with the various accounts of the nature of fictional objects make a difference to the answers we end up giving to the questions we are ultimately interested in—questions about the nature of explanation, say, or about how models are evaluated in light of the data we have collected. If we grant that understanding how modeling works is of only derivative interest, then because I have not shown that different answers to the nature question *will* make a difference of that sort, my arguments to this point do perhaps fail to establish that bracketing the nature question is not an option. To put it another way, the objection we are considering does pose a threat to my attempt to argue that the autonomy thesis (step [4] in the legitimization argument) and claim that we can bracket (step [5]) are false. But the objection does nothing to undercut my attempt to show that the autonomy thesis and the claim that we can bracket do not follow from the premises of the legitimization argument. Indeed, as the objection allows for the possibility that in the situation at hand the answer to the nature question will make a difference to the answers we give to the questions we are centrally concerned with as philosophers of science, it allows that the conclusions of the legitimization argument do not follow from its premises. Granting even this much, I would maintain, puts the bracketer in an uncomfortable position. When evaluating a core account which engages in the as-if practice for *X*'s, it is prima facie entirely reasonable to ask, as part of the evaluation, whether there are indeed *X*'s, and if so, whether they are the right sort of thing to play the roles the core account would seem to require of them. If the legitimization argument fails, the bracketer has more work to do to support her claim that such questions can be brushed aside.

If the objection succeeds, then, it stands in the way of only the second of my two main claims: that the bracketing strategy is unavailable to us when we offer core accounts which engage in the as-if practice for *X*'s (or, to put it in other words, that the existence and nature questions about *X*'s cannot be brushed aside in such a situation). But I do not think the objection succeeds even in this direction. To see this, go back to the argument against the possibility of bracketing the existence question. If we take no stand on whether there are fictional objects, then even if we know somehow that (S) says something true, it is hard to see what we could come to understand about how modeling works by invoking (S). Compare this with the situation of someone who speaks not a word of some language, but who knows, somehow, that a certain sentence of it makes a true claim; clearly such a person could not gain any understanding whatever of the topic of the sentence by considering that sentence. I would not claim that the two situations are the same, of course; but it does seem to me that they are far too close for the comfort of the would-be bracketer. Now, the objection we are considering does nothing to cast doubt on this part of my argument. But then there are two points to be made. First, even if my argument succeeds only in showing that the existence question about *X*'s cannot be bracketed when we are developing and evaluating core accounts which engage in the as-if practice for *X*'s, that is enough to bring back into the fold just the sort of metaphysical issue the bracketer was hoping to put aside. Second, once we grant that we cannot put aside the existence question about *X*'s, it will become very difficult to keep the nature question at bay, for the simple reason that deciding whether there are metaphysically controversial *X*'s typically involves thinking about what sorts of things *X*'s might be if there are any. Thus it seems that both the spirit and the letter of my second main claim survive the objection in question.

5. CLOSING POLEMICS

I have argued that the legitimization argument for the autonomy thesis and the corresponding bracketing strategy fails. In doing so, I have tried to establish, moreover, that the autonomy thesis is false, and that the bracketing strategy is not an option. I am arguing, that is, that if we wish to achieve the aims we have *as philosophers of science*, which include developing satisfying accounts of scientific representation, modeling, explanation, idealization, confirmation, and the like, and we seek to do so by engaging in talk of models, mathematical structures, or fictional objects, as many do, then we will have to address metaphysical questions about the existence and nature of such creatures.[40]

Let me close with an analogy to underscore the point, and to up the polemical ante on the way out. Suppose we are philosophers of mind trying to come up with an account of the relationship between the mental and the physical which explains the correlations we observe between mental and physical events. Suppose then that one of us says, "Here's my account: God is constantly intervening in our mental lives to produce pain sensations in us when we're kicked, sensations of blueness when we're standing in front of blue objects, and so on; God is also constantly intervening in the physical world to make sure that my leg moves when I decide to kick something, and so on." A little old-fashioned, but fair enough. Suppose, however, that our colleague then goes on to say, "I take no stand on whether there is any such thing as God [the existence question], nor on whether, if there is such a thing, it's the kind of thing that would take any interest in human affairs [the nature question]. Those are questions for metaphysics and the philosophy of religion; I'm just going to stick to the philosophy of mind. And my account of the question before us as philosophers of mind is the account I gave a moment ago." Such a philosopher would be employing the bracketing strategy; and it clearly wouldn't do.

ACKNOWLEDGMENTS

Thanks to Kyle Stanford and Zanja Yudell for their incisive comments on an earlier version of this paper. I doubt that I have addressed all their concerns in ways that they will find satisfying, but I hope they at least think the paper better for their efforts. Thanks also to Paul Teller for conversation and correspondence which resulted in a clearer formulation of my target in the early stages of my thinking, to Roman Frigg, Ron Giere, Peter Godfrey-Smith, and Arnon Levy for multiple conversations on these issues, and to Anjan Chakravartty for helpful correspondence.

REFERENCES

Cartwright, Nancy. 1983. *How the Laws of Physics Lie*. Oxford: Clarendon Press.
Cartwright, Nancy. 1989. *Nature's Capacities and their Measurement*. Oxford: Clarendon Press.

[40] See section 2 for more about what I do and do not mean here by "address."

Contessa, Gabriele. 2010. "Scientific Models and Fictional Objects." *Synthese* 172: 215–29.

Frigg, Roman. 2010. "Models and Fiction." *Synthese* 172: 251–68.

Frisch, Mathias. 1998. "*Theories, Models, and Explanation*." PhD diss., University of California, Berkeley.

Godfrey-Smith, Peter. 2006. "The Strategy of Model-Based Science." *Biology and Philosophy* 21: 725–40.

Lewis, David. 1978. "Truth in Fiction." *American Philosophical Quarterly* 15: 37–46.

Lewis, David. 1986. *On the Plurality of Worlds*. Oxford: Basil Blackwell.

Rosen, Gideon. 1994. "What is Constructive Empiricism?" *Philosophical Studies* 74: 143–78.

Thomasson, Amie L. 1999. *Fiction and Metaphysics*. Cambridge: Cambridge University Press.

Thomasson, Amie L. 2003. "Speaking of Fictional Characters," *Dialectica* 57: 205–23.

Thomson-Jones, Martin. 1997. "Models and the Semantic View." PhilSci Archive, University of Pittsburgh. http://philsci-archive.pitt.edu/id/eprint/8994.

Thomson-Jones, Martin. 2007. "Missing Systems and the Face Value Practice." PhilSci Archive, University of Pittsburgh. http://philsci-archive.pitt.edu/id/eprint/3519.

Thomson-Jones, Martin. 2010. "Missing Systems and the Face Value Practice." *Synthese* 172: 283–99.

Thomson-Jones, Martin. 2012. "Modeling without Mathematics." *Philosophy of Science* 79: 761–72.

Thomson-Jones, Martin. Forthcoming. "Realism about Missing Systems." In *The Scientific Imagination, edited by* P. Godfrey-Smith and A. Levy, Oxford: Oxford University Press.

Toon, Adam. 2012. *Models as Make-Believe: Imagination, Fiction and Scientific Representation*. New York: Palgrave MacMillan.

van Fraassen, Bas C. 1980. *The Scientific Image*. Oxford: Clarendon Press.

van Fraassen, Bas C. 1994. "Rosen on Constructive Empiricism." *Philosophical Studies* 74: 179–192.

van Fraassen, Bas C. 2005. "On Taking Stances: An Interview with Bas van Fraassen." By K. Walden. *The Harvard Review of Philosophy* 13 (2): 86–102.

van Fraassen, Bas C. 2008. *Scientific Representation: Paradoxes of Perspective*. Oxford: Oxford University Press.

Wallace, David (2011). "Taking particle physics seriously: A critique of the algebraic approach to quantum field theory." *Studies in History and Philosophy of Science Part B* 42 (2):116–125.

Weisberg, Michael. 2013. *Simulation and Similarity: Using Models to Understand the World*. Oxford: Oxford University Press.

INDEX